# PRINCIPLES OF PESTICIDE CHEMISTRY

**S. K. HANDA**
*Head*
*Division of Agricultural Chemicals*
*Indian Agricultural Research Institute*
*New Delhi*

AGROBIOS (INDIA)

*Published by:*
**AGROBIOS (INDIA)**
Agro House, Behind Nasrani Cinema
Chopasani Road, Jodhpur 342003
Phone: 91-0291-2643993
E. mail: agrobiosindia@gmail.com, info@agrobiosindia.com
Website: www.agrobiosindia.com

AGROBIOS (INDIA)



**ISBN:** 978-81-7754-216-5

***Published by:*** Dr. Updesh Purohit for Agrobios (India), Jodhpur
***Lasertypeset at:*** Yashee Computers and Printers, Jodhpur
***Cover Design by:*** Reena
***Printed at:*** Agrobios Digitals, Jodhpur

# Preface

The present book on Principles of Pesticides Chemistry is the outcome of difficulties encountered by the author and his colleagues while imparting knowledge of agricultural chemicals to their students. Several excellent books of pesticides have been published in the last decade but none of them could be recommended as a text to our students for the simple reason they fail to cover all aspects of prescribed syllabus of Agricultural Universities of our country. An attempt has been made to give an over view of all chemicals aspects of plant protection. The book on Principles of Pesticides Chemistry presents a detailed characteristic of insecticides, fungicides, nematicides, fumigants, acaricides, herbicides, attractants, repellents and pheromones. Information is also given on Insecticides Act, biological control of pesticides, toxicology, formulation and residues aspect of pesticides.

I am very much aware of certain limits of the book due to various difficulties in printing which will be eliminated in the next edition. Finally, I wish to place on record my appreciation of help and encouragement given by my colleagues in the Department of Agricultural Chemicals, Indian Agricultural Research Institute, New Delhi.

In the end, I request the readers to offer their suggestions for further improvement of the book and the same will be gratefully acknowledged and incorporated in the next edition.

**(S.K. Handa)**

# CONTENTS

# CHAPTER 1

# Introduction

Pesticide means any substance intended for preventing, destroying, attracting, repelling or controlling any pest including unwanted species of plants or animals during production, storage, transport, distribution and processing of food, agricultural commodities, or animal feeds or which may be administered to the animals for the control of ectoparasites. The term includes substances intended for use as a plant growth regulator, defoliant, desiccant, fruit thinning agent or sprouting inhibitor and substances applied to the crops either before or after harvest to protect the commodity from deterioration during storage and transport. The term normally excludes fertilizers, plant and animal nutrients, food additives and animal drugs. Pesticide is a general term used in connection with any pest. More specific terms are available for individual pest categories. Pests are any organisms, insects, rodents, nematodes, fungal diseases, weeds, herbs, bacteria etc. which are injurious to the health of man or to man's economic efforts.

1. **Insecticides:** are substances that prevent, destroy kill mitigate insects (belonging to class insects, Phylum athropoda).
2. **Fungicides:** are substances that prevent, destroy or inhibit the growth of fungi/diseases in crops.
3. **Herbicides:** are substances which are used for preventing or inhibiting growth of plants or for killing weeds.
4. **Rodenticides:** are substances that inhibit growth, destroy or kill rodents (Class Mammalia, order Rodentia)
5. **Nematicides:** are substances that prevent, destroy, repel or inhibit the nematodes.
6. **Chemosterilants:** are substances that sterlize the pests.
7. **Molluscides:** are substances that prevent, repel, destroy or inhibit the growth of the Phylum Mollusca.
8. **Plant growth Regulators:** are substances that cause the acceleration or retardation of the rate of growth or rate of maturation.
9. **Defoliants:** are substances that cause the plant leaves to die and fall away.
10. **Desiccants:** are substances that cause to drain moisture out of the plants causing them to dry.
11. **Attractants:** are substances that attract the pests.

12. **Repellents:** are substances that repels the pest from a treated plant.

## Ideal Pesticide

It should have the following properties:

1. It should have a broad spectrum to kill the pests.
2. It should be non-toxic to mammals.
3. It should be non toxic to plants.
4. It should have low production cost.
5. It should be easily formulated.
6. It should have optimal residue persistence.
7. It should have low toxicity to beneficial insects.
8. It should be compatible with other pesticides.
9. It should not produce off taste on vegetables on which it is sprayed.
10. It should be non inflammable and should have non corrosive action on metals.

## Steps Involved in Commercial Development of Pesticide Starting from Laboratory Synthesis

1. Synthesis
2. The preliminary screening
3. Formulation
4. Toxicity studies, (acute, oral, dermal etc.)
5. Metabolism studies
6. Residue studies
7. Registration
8. Promotion
9. Pilot plant studies
10. Commercial Development

## Reasons for Use of Pesticides

1. Pesticides are used to ensure better protection at harvest against unpredictable losses caused by plant diseases and pests.
2. Pesticides are used to improve both quality and quantity of food.
3. Pesticides are used to decrease the extent of vector born and other diseases in humans and animals

## Production of Pesticides in Organised Sector in India

Of the total pesticides used in India, 80 per cent are insecticides, 11 per cent are fungicides, and 7 per cent are herbicides. In India, total production of pesticides has increased from 250 metric tonnes in 1947-48 to 80,000 metric tonnes during the year 1996-97. India has installed capacity of infrastructure for manufacture of 62 technical grade pesticides (Table 1).

**TABLE 1:** Indian production of major technical grade pesticides

| Product | Production (1000 tonnes) | | | % of installed capacity |
|---|---|---|---|---|
| | 1995-96 | % change | 1996-97 | |
| **Insecticides** | | | | |
| Acephate | 0.5 | 0 | 0.5 | 100.0 |
| BHC | 30.2 | +1.0 | 30.5 | 82.4 |
| Chlorpyrifos | 0.2 | 0 | 0.2 | 25.0 |
| Cypermethrin | 1.0 | +10.0 | 1.1 | 100.0 |
| DDT | 6.0 | -16.7 | 5.0 | 54.9 |
| Dichlorvos | 1.6 | 0 | 1.6 | 61.5 |
| Dimethoate | 1.7 | 0 | 1.7 | 73.9 |
| Endosulfan | 7.4 | -5.4 | 7.0 | 95.9 |
| Ethion | 0.8 | 0 | 0.8 | 88.9 |
| Fenitrothion | 0.03 | -33.3 | 0.02 | 4.0 |
| Fenthion | 0.2 | 0 | 0.2 | na |
| Fenvalerate | 1.2 | +16.7 | 1.4 | 100.0 |
| Malathion | 2.9 | +3.4 | 3.0 | 39.5 |
| Metasystox | 0.4 | +50.0 | 0.6 | 100.0 |
| Monocrotophos | 7.8 | +7.7 | 8.4 | 81.6 |
| Parathion-methyl | 2.4 | 0 | 2.4 | 53.3 |
| Phorate | 3.3 | +15.2 | 3.8 | 97.4 |
| Phosalone | 0.4 | 0 | 0.4 | 40.0 |
| Phosphamidon | 2.0 | +5.0 | 2.1 | 65.6 |
| Quinalphos | 2.5 | +20.0 | 3.0 | 100.0 |
| Temephos | 0.3 | 0 | 0.3 | na |
| **Fungicides** | | | | |
| Captan/ captafol | 0.5 | 0 | 0.5 | 83.3 |
| Carbendazim | 0.7 | -14.3 | 0.6 | 50.0 |
| Mancozeb | 4.0 | -5.0 | 3.8 | 95.0 |
| Thiram | 0.02 | 0 | 0.02 | 10.0 |
| Tridemorph | 0.2 | 0 | 0.2 | 40.0 |
| Ziram | 0.1 | 0 | 0.1 | 25.0 |
| **Herbicides** | | | | |
| 2,4-D | 0.9 | +11.1 | 1.0 | 55.6 |
| Butachlor | 1.2 | -33.3 | 0.8 | 17.8 |
| Fluchloralin | 0.14 | -35.7 | 0.09 | 30.0 |
| Isoproturon | 2.7 | +11.1 | 3.0 | 96.8 |
| **Fumigants** | | | | |
| Aluminium | 1.5 | -6.7 | 1.4 | 100.0 |

| Product | Production (1000 tonnes) | | | % of installed capacity |
|---|---|---|---|---|
| | 1995-96 | % change | 1996-97 | |
| Phosphide | | | | |
| Methyl bromide | 0.08 | 0 | 0.08 | 26.7 |
| **Total** | **85.5** | **+0.4** | **85.8** | **73.6** |

## LEVEL OF PESTICIDES CONSUMPTION IN VARIOUS COUNTRIES

The consumption of pesticides kg/ha in developing countries is low as compared to the developed countries. India use only 0.3 kg/ha of pesticides while in Korea and Japan it is 6.60 and 10.80 kg/ha respectively (Table 2).

**TABLE 2:** Pesticide used on major crops in some countries

| Country | Pesticide used (Kg/ha) | Country | Pesticide used (Kg/ha) |
|---|---|---|---|
| Japan | 10.80 | Indonesia | 0.60 |
| Korea | 6.60 | India | 0.30 |
| Europe | 1.90 | Latin America | 6.22 |
| USA | 1.50 | Oceanic | 0.20 |
| Thailand | 1.40 | Africa | 0.13 |

## PRESENT CONSUMPTION AND FUTURE DEMAND

The planning commission's working group on Pesticide Industry has placed the demand for technical grade pesticides at 107, 864 metric tonnes for the year 1994-95 (Table 3). The production and consumption of pesticides have not witnessed an increase at the desired pace in spite of the fact that the area under plant protection has been increasing considerably year after year.

**TABLE 3:** Demand pattern during VIIIth five year plan

| Group of pesticides | (Tonnes) Demand as proposed by working group | | |
|---|---|---|---|
| | 1989-90 | 1991-92 | 1994-95 |
| Fungicides | 11,387 | 13,858 | 15,564 |
| Weedicides | 5,710 | 6,541 | 8,125 |
| Rodenticides & Fumigants | 1,790 | 2,062 | 2,615 |
| Acaricides | 145 | 154 | 170 |
| **Total** | **79,482** | **91,040** | **1,07,864** |

***Source:*** Report of the working group on Pesticide Industry (VIIIth Five Year Plan), November 1989, Government of India, Cited from David B.V. (ed) (1992) Pest Management and Pesticides: Indian Scenario, Namrutha Publications, Madras, p. 250.

## CROPWISE CONSUMPTION OF PESTICIDES

The figures of cropwise consumption indicate that cotton claims most of the pesticides 44.5% while for paddy it is 22% of the total pesticide used followed by jowar (8.9%) and wheat (6.4%). The consumption of pesticide on fruits and vegetables is only 7% (Table 4)

**TABLE 4:** Cropwise consumption of pesticides

| Crop | Estimated pesticide consumption (in million Rs.) | % to total pesticides consumption |
|---|---|---|
| Paddy | 1272.05 | 22.0 |
| Wheat | 354.18 | 6.4 |
| Jowar | 495.40 | 8.9 |
| Bajra | 95.05 | 1.8 |
| Barley | 0.00 | 0.0 |
| Ragi | 24.04 | 0.4 |
| Maize | 63.62 | 1.1 |
| Arhar | 155.20 | 2.8 |
| Gram | 12.20 | 0.2 |
| Jute | 15.38 | 0.3 |
| Cotton | 2462.13 | 44.5 |
| Sugar cane | 38.41 | 0.7 |
| Groundnut | 136.84 | 2.5 |
| Rape and mustard | 9.10 | 0.2 |
| Soybean | 0.95 | 0.0 |
| Sunflower | 3.04 | 0.1 |
| Onion | 10.92 | 0.2 |
| Potato | 2.45 | 0.0 |
| Tobacco | 3.93 | 0.1 |
| Tapioca | 0.10 | 0.0 |
| Fruits and vegetables (excluding Onion and Potato) | 389.38 | 7.0 |
| **Total** | **5556.37** | **100.0** |

## STATEWISE CONSUMPTION OF PESTICIDES IN AGRICULTURE

Amongst the states in terms of highest consumption of pesticides, Tamil Nadu is highest with 16.5%. The share of Andhra Pradesh, Uttar Pradesh, Maharashtra in total consumption of pesticides is about 13.0, 14.0 and 11.24 per cent respectively. Other states account for less than nearly 7 per cent, but most of the states account for less than 3 per cent (Table 5).

**TABLE 5:** Consumption of pesticides in various states/union territories

| Sl.No. | State/Union Territories | Consumption (Metric tonnes) |
|---|---|---|
| 1. | Andhra Pradesh | 9,910 |
| 2. | Arunachal Pradesh | 30 |
| 3. | Assam | 595 |
| 4. | Bihar | 1,700 |
| 5. | Gujarat | 5,500 |
| 6. | Goa | 22 |
| 7. | Haryana | 4,500 |
| 8. | Himachal Pradesh | 718 |
| 9. | Jammu & Kashmir | 110 |
| 10. | Karnataka | 3,900 |
| 11. | Kerala | 1,100 |
| 12. | Madhya Pradesh | 4,500 |
| 13. | Maharashtra | 6,020 |
| 14. | Manipur | 50 |
| 15. | Meghalaya | 45 |
| 16. | Mizoram | 15 |
| 17. | Nagaland | 12 |
| 18. | Orissa | 1,800 |
| 19. | Punjab | 5,770 |
| 20. | Rajasthan | 2,758 |
| 21. | Sikkim | 20 |
| 22. | Tamil Nadu | 12,500 |
| 23. | Tripura | 164 |
| 24. | Uttar Pradesh | 8,480 |
| 25. | West Bengal | 5,000 |
| 26. | Andaman & Nikobar | - |
| 27. | Chandigarh | - |
| 28. | Delhi | 60 |
| 29. | Dadra & Nagar Haveli | 4 |
| 30. | Daman & Diu | - |
| 31. | Pondicherry | 135 |
| 32. | Lakshadweep | 0.70 |
| | **Total** | **75,417.70** |

## PESTICIDES REGISTERED, NOT APPROVED FOR USE, BANNED, RESTRICTED FOR USE AND UNDER REVIEW

The list of pesticides registered, not approved for use, banned, restricted for use and under review are given in Tables 6, 7, 8, 9 and 10.

**TABLE 6:** List of insecticides registered on regular basis under section 9 (3) of the insecticides act, 1968

| Name of the Insecticied | |
|---|---|
| 1. Acephate | 2. Alachlor |
| 3. Aldicarb | 4. Allethrin |
| 5. Alphanapthyl Acetic Acid | 6. Aluminium phosphide |
| 7. Anilophos | 8. Atrazine |
| 9. Alphacypermethrin | 10. Aureofungin |
| 11. Barium Carbonate | 12. Benzene Hexachloride (BHC) |
| 13. Benthiocarb (Thiobencarb) | 14. Benomyl |
| 15. Bitertanol | 16. Bromadiolone |
| 17. Butachlor | 18. Calcium Cyanide |
| 19. Captafol | 20. Captan |
| 21. Cartap Hydrochloride | 22. Carbaryl |
| 23. Carbendazim | 24. Carbofuran |
| 25. Carboxin | 26. Chlorbenzilate |
| 27. Chlorfenvinphos | 28. Chlormequat Chloride (CCC) |
| 29. Chlorothalonil | 30. Chlorpyriphos |
| 31. Copper Oxychloride | 32. Copper Sulphate |
| 33. Coumachlor | 34. Coumatetralyl |
| 35. Cuprous Chloride | 36. Cypermethrin |
| 37. Cyfluthrin | 38. Dalapon |
| 39. Decamethrin (Deltamethrin) | 40. Dichloro Diphenyl Trichloroethane (DDT) |
| 41. Dichloropropane and Dichloropropanes mixture (DD mixture) | 42. Dichlorvos (DDVP) |
| 43. Dicofol | 44. Diflubenzuron |
| 45. Dimethoate | 46. Dinacap |
| 47. Dithianon | 48. Diuron |
| 49. Dodine | 50. Diazinon |
| 51. Dieldrin (Restricted for Locust Control by PPA) | 52. 2,4-Dichlorophenoxy Acetic Acid (2,4-D, Sodium Amine and Ester salts) |
| 53. Ethylene Dibromide and Carbon Tetrachloride mixture (EDCT mixture 3:1) | 54. Edifenphos |
| 55. Endosulfan | 56. Ethofenprox (Etofenprox) |
| 57. Ethephon | 58. Ethylene Dibromide (EDB) |
| 59. Ethyl Mercury Chloride | 60. Ethion |
| 61. Fenitrothion | 62. Fenarimol |
| 63. Fenobucarb (EPMC) | 64. Fenthion |
| 65. Fenvalerate | 66. Ferbam |
| 67. Fluchloralin | 68. Fluvalinate |

| Name of the Insecticied | |
|---|---|
| 69. Formothion | 70. Fosetyl-Al |
| 71. Gibberellic Acid | 72. Glyphosate |
| 73. Iprodione | 74. Isoproturon |
| 75. Kitazin | 76. Lime Sulphur |
| 77. Lindane (Gamma BHC) | 78. Malathion |
| 79. Maleic Hydrazide (MH) | 80. Mancozeb |
| 81. Menazon | 82. Metaldehyde |
| 83. Methabenzthiazuron | 84. Methoxy Ethyl Mercury Chloride (MEMC) |
| 85. Methyl Bromide | 86. Methyl Bromide + Ethylene dibromide (1:1 & 3:1) |
| 87. Methyl Chlorophenoxy Acetic Acid (MCPA) | 88. Metalaxyl |
| 89. Metoxuron | 90. Methomyl |
| 91. Metribuzon | 92. Metolachlor |
| 93. Monocrotophos | 94. Mono Sodium Methane Arsonate (MSMA) |
| 95. Myclobutanil | 96. Methyl Parathion |
| 97. Nickle Chloride | 98. Nicotine Sulphate ** |
| 99. Oxadiazon | 100. Oxycarboxin |
| 101. Oxydemeton-methyl | 102. Oxyfluorfen |
| 103. Para dichlorobenzene | 104. Paris Green (Copper Acetoarsenite) |
| 105. Paraquat dichloride | 106. Pendimethalin |
| 107. Permethrin | 108. Phenthoate |
| 109. Phenyl Mercury Acetate (PMA)** | 110. Phorate |
| 111. Phosalone | 112. Phosphamidon |
| 113. Pirimiphos-methyl | 114. Profenofos |
| 115. Propanil | 116. Propetamphos |
| 117. Propoxur | 118. Pyrethrum |
| 119. Quinalphos | 120. Sevidol (Carbaryl and Gamma BHC) |
| 121. Simazine | 122. Sirmate |
| 123. Sodium Cyanide | 124. Streptocycline (Streptomycin) |
| 125. Sulphur | 126. Temephos |
| 127. Trichloro Acetic Acid (TCA) | 128. Triallate |
| 129. Trichlorphom | 130. Tricyclazole |
| 131. Tridemorph | 132. Trifluralin |
| 133. Thiometon | 134. Thiram |
| 135. Thiophanate Methyl | 136. Triazophos |
| 137. Triadimefon | 138. Warfarin |
| 139. Zinc Phosphide | 140. Zineb |
| 141. Ziram | 142. Hexaconazole |

| Name of the Insecticied | |
|---|---|
| 143. Pretilachlor | 144. Propiconazole |

** for export only.

**TABLE 7:** List of pesticides not approved for use

| Name of Pesticides | |
|---|---|
| 1. Calcium aresenate | 2. EPN |
| 3. Azinphos-methyl | 4. Lead Arsonate |
| 5. Mevinphos (Phosdrin) | 6. Carbophenothion |
| 7. Vamidothion | 8. Mephosfolen |
| 9. Azinphos Ethyl | 10. Binapacryl |
| 11. Dicrotophos | 12. Thiodmeton/Disulfoton |
| 13. Fention Acetate | 14. Fention Hydroxide |
| 15. Chinomethionate (Morestan) | 16. Ammonium Sulphamate |
| 17. Leptaphos (Phosvel) | |

**TABLE 8:** List of banned pesticides

| Name of Pesticides | |
|---|---|
| 1. Dibromochloropropane (DBCP) | 2. Endrin |
| 3. Pentachlornitrobenzene (PCNB) | 4. Pentachlorophenol (PCP) |
| 5. Toxaphene | 6. Ethyl Parathion |
| 7. Chlordane | 8. Heptachlor |
| 9. Aldrin (w.e.f. 1.1.1994) | 10. Paraquat-di-methyl Sulphate |
| 11. Nitrofen | 12. BHC |
| 13. Tetradifen | |

**TABLE 9:** List of pesticides restricted for use in India

| Sl. No. Name of Pesticide | Type of restriction(s) imposed) |
|---|---|
| Aluminium Phosphide. | – It is to be sold only to Government Undertaking/Organisations and to be used under the strict supervision of Government Expert of Pest Control Operations whose expertise is approved by the plant protection Adviser to the Government of India. |
| B.H.C. | – The use of BHC on vegetabales, fruits, oilseed crops and preservation of foodgrains is baned. |
| Chlorbenzilate. | – Use in agriculture is banned. If required it can be imported by Govt. /Semi. Govt. Organisations for pre folbex strips for making it available to bee keepers for controlling mites of honeybees. |
| Captafol. | – Shall be used only as seed dresser. Its use as foliar spray is banned. |
| D.D.T. | – Use of DDT in agriculture is banned. Use of DDT for the public health programme upto 10,000 MTs per annum is restricted. |

| Sl. No. Name of Pesticide | Type of restriction(s) imposed) |
|---|---|
| Dieldrin. | – Use of Dieldrin shall be restricted for locust control in desert areas by plant protection Adviser to the Government of India. |
| Ethylene dibromide (EDB). | – Use of EDB shall be restricted as a fumigant for foodgrains through Central Government, State Government, State Government Undertakings, Government Organisations such as Food Corporation of India, Central Warehousing Corporations, State Warehousing Corporation and Pest Control Operators whose expertise is approved by Plant Protection Adviser to the Govt. of India. |
| Methyl Bromide | – Methyl Bromide is to be used as a fumigant in the manner in which Aluminium Phosphide is used. |
| Sodium Cyanide | – Use of Sodium Cyanide shall be restricted for fumigation of cotton bales by Plant Protection Adviser to Govt. of India under Expert supervision. |
| Phenyl mercury acetate (PMA) | – Use of PMA is prohibited in India w.e.f. 1.1.73 but it may continue to be produced in the country only for export. Nonconsumed stock as on 1.1.73 shall be used for Export purpose only and shall be subject to the modification in the Certificate of registration to this effect. |
| Lindane | – Use of Lindane formulation generating smoke for indoor use is prohibited in India provided that lindane may be continued to be produced in the country for use for control of insect pests of field crops and subject to the modification in the Certificate of registration to this effect. |
| Nicotine Sulphate | – Use of Nicotine Sulphate is prohibited in India provided that may be continued to be produced in the country only for Export purpose and subject to the modification in the Certificate of Registration to this effect. |

**TABLE 10:** List of pesticides which are under review by government of India.

| Name of Pesticides | |
|---|---|
| 1. Alachlor | 2. Benomyl |
| 3. Copper acetoarsenite | 4. Diuron |
| 5. Ethyl Mercury Chloride | 6. Fenarimol |
| 7. Menazon | 8. Methomyl |
| 9. Oxyflourfen | 10. Sodium Methane Arsonate |
| 11. Calcium Cyanide | 12. Phosphamidon |
| 13. Thiometon | 14. Triazophos |
| 15. Tridemorph | 16. Monocrotophos |
| 17. Ziram | 18. Zineb |

## LIST OF QUALITY CONTROL LABORATORIES IN INDIA

In India the number of pesticide testing laboratories in states/union territories are 40 and capacity per annum for testing of samples for quality control is 45,556 (Table 11)

**TABLE 11:** Pesticides testing laboratories in state/union territory.

| State/Union Territory | No. of labs | Location | Capacity annum |
|---|---|---|---|
| Andhra Pradesh | 4 | Rajendranagar, Guntur Anantapur & Tadepalligudem | 8,520 |
| Assam | 1 | Guwahati | 166 |
| Bihar | 1 | Patna | 500 |
| Gujarat | 2 | Junagarh & Gandhinagar | 2,000 |
| Jammu & Kashmir | 1 | Srinagar | 600 |
| Karnataka | 2 | Bangalore & Bellary | 2,400 |
| Kerala | 1 | Trivandrum | 2,000 |
| Madhya Pradesh | 1 | Jabalpur | 1,000 |
| Maharashtra | 4 | Pune, Amravati, Thane & Aurangabad | 3,840 |
| Manipur | 1 | Mantripukhri | 30 |
| Mizoram | 2 | Aizwal | - |
| Orissa | 1 | Bhubaneswar | 1,000 |
| Punjab | 3 | Ludhiana, Bhatinda & Amritsar | 900 |
| Rajasthan | 2 | Durgapura (Jaipur) & Bikaner | – |
| Tamil Nadu | 9 | Coimbatore, Kovilpatti, Madurai, Trichurapalli, Aduthurai, Salem Erode, Kanchepuram & Cuddalore | 16,200 |
| Uttar Pradesh | 3 | Meerut, Lucknow & Varanasi | 3,000 |
| West Bengal | 1 | Midnapore | 450 |
| Pondicherry | 1 | Pondicherry | 450 |
| **Total** | **40** | | **45,556** |

# CHAPTER 2

# History of Pesticides Development

The history of pesticide development is given in table 1 and table 2.

**TABLE 12:** History of pesticide development

| Name | Class | Year |
|---|---|---|
| Red Squil | Rodenticide | BC |
| Sulphur | Fumigant | 1000 |
| Tobacco | Insecticide | 1690 |
| Nicotine | Fumigant | 1773 |
| Soap | Insecticide | 1787 |
| Turpentine emulsion | Repellant | 1787 |
| Fish oil | Insecticide | 1820 |
| Sulphur | Fungicide | 1821 |
| Mercuric chloride and alcohol | Insecticide | 1822 |
| Phosphorus | Rodenticide | 1845 |
| Carbon disulphide | Fumigant | 1854 |
| Pyrethrum | Insecticide | 1858 |
| Paris green | Insecticide | 1867 |
| Kerosene | Insecticide | 1868 |
| Hydrogen cyanide | Fumigant | 1877 |
| Hydrate lime and arsenicals | Insecticide | 1890 |

**TABLE 13:** Discovery of major commercial pesticides

| Name | Class | Year |
|---|---|---|
| DDT | Insecticide | 1939 |
| Schradan | Insecticide | 1941 |
| BHC and Lindane | Insecticide | 1942 |
| 2,4-D | Herbicide | 1943 |
| MCPA | Herbicide | 1943 |
| Warfarin | Rodenticide | 1944 |
| Chlorodane | Insecticide | 1945 |
| Aldrin, Dieldrin | Insecticide | 1945 |
| Hepltachlor | Insecticide | 1945 |
| Toxaphene | Insecticide | 1948 |

| Name | Class | Year |
|---|---|---|
| Allethrin | Insecticide | 1949 |
| Malathion | Insecticide | 1950 |
| Captan | Fungicide | 1951 |
| Dichlorvas | Insecticide | 1951 |
| Diazinon | Insecticide | 1952 |
| Endrin | Insecticide | 1952 |
| Steptomycin | Fungicide | 1952 |
| Atrazine and Simazine | Herbicide | 1952 |
| Phorate | Insecticide | 1954 |
| Dimethoate | Insecticide | 1956 |
| Phosphamidon | Insecticide | 1956 |
| Carbaryl | Insecticide | 1956 |
| Paraquat | Herbicide | 1958 |
| Fenitrothion | Insecticide | 1959 |
| Propoxur | Insecticide | 1962 |
| Chlorfenvenphos | Insecticide | 1963 |
| Monocrotophos | Insecticide | 1965 |
| Dimethirimol | Fungicide | 1965 |
| Quinalphos | Insecticide | 1965 |
| Aldicarb | Insecticide | 1966 |
| Carbofuran | Insecticide | 1967 |
| Benomyl | Fungicide | 1967 |
| Acephate | Insecticide | 1969 |
| Primiphosmethyl | Insecticide | 1970 |
| Fenvalerate | Insecticide | 1976 |
| Permethrin | Insecticide | 1977 |
| Cypermethrin | Insecticide | 1977 |
| Deltamethrin | Insecticide | 1977 |
| Fenpropathrin | Insecticide | 1980 |
| Cyfluthrin | Insecticide | 1980 |
| Sulphonylureas | Herbicide | 1982 |
| Fluvalinate | Insecticide | 1983 |
| Alphamethrin | Insecticide | 1983 |
| Ethofenprox | Insecticide | 1986 |
| Esfenvalerate | Insecticide | 1986 |
| Talomethrin | Insecticide | 1986 |
| Tefluthrin | Insecticide | 1987 |

# CHAPTER 3

# Innovation in Pesticides Chemistry

The following methods are used for the development of new pesticides.

1. Random screening
2. All round screening
3. Models of natural compounds
4. Chemical modifications
5. Mathematical models.

## RANDOM SCREENING

Random screening is the biological evaluation of chemical compounds with no structural relation to any known pesticides. This was the earlier method at the outset of insecticide research and DDT, HCH are important examples of random screening.

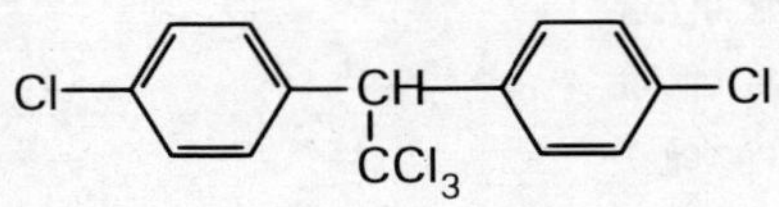

DDT

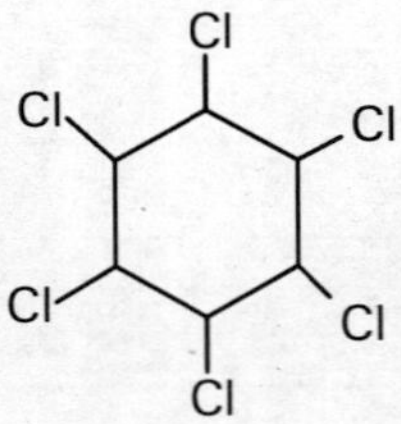

HCH

## ALL ROUND SCREENING

All round screening takes in only those compounds which have been synthesised for a specific purpose. The biological screening however is done in all major fields and is not restricted to expected activity. The important example is synthesis of diflubenzuron. The structure of dichlobenzil was introduced in 1960 and the structure was modified in the hope of improving certain properties as an herbicide. An extensive synthesis programme finally led to diflubenzuron. The compound inhibits the chitin synthesis specific to invertebrates and is considered as to be a selective insecticide.

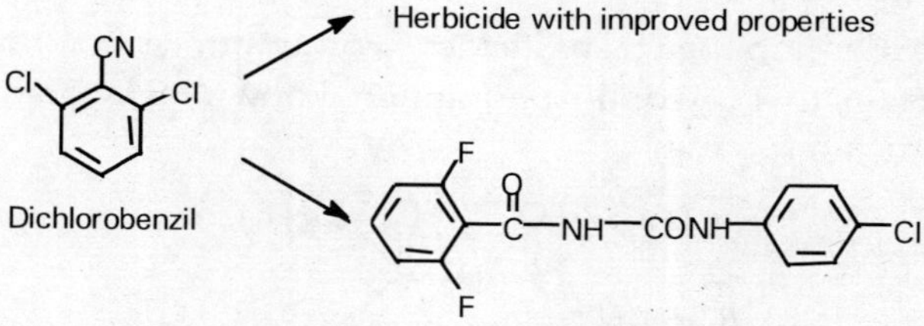


Dichlorobenzil

Diflubenzuron

## MODELS OF NATURAL COMPOUNDS

The crude extract of the botanicals are screened for their biological properties. If the crude extract is found to be toxic, then the compounds are isolated and their structures are elucidated. Based on the structures of these compounds, new pesticides are synthesised. For example, pyrethrum is obtained from dried flower heads of *Chrysanthemum cinerariefolium* by extraction with kerosene. It contains six insecticidal components which are collectively termed as pyrethrins. Based on the structures of pyrethrins, synthetic pyrethroids like fenvalerate, cypermethrin, and deltamethrin were synthesised. (For structures of natural pyrethrines and synthetic pyrethroids, consult chapter of botanicals).

Another example of development of synthetic pesticides based on the model of botanical are carbamates. Phostgmine was isolated from calabar beans of poisonous plant *Physostigma venenosa*. In 1925, Stedmon and Berger established the structure of physostigmine and it was the first known carbamate ester from natural product. Based on the structure of phostigmine several carbamates like carbaryl, carbofuran, and aldicarb were synthesised.

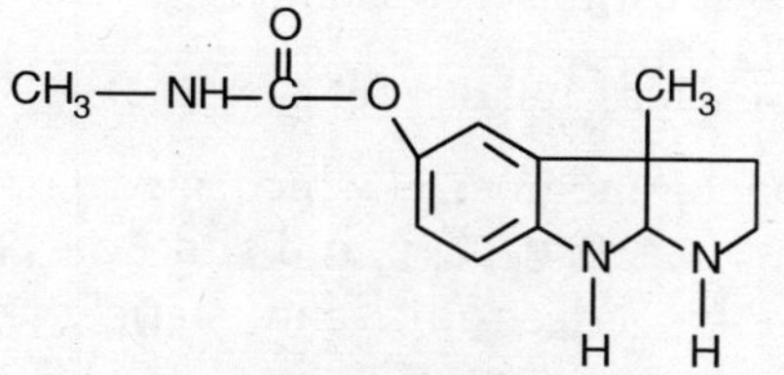

Physostigmine

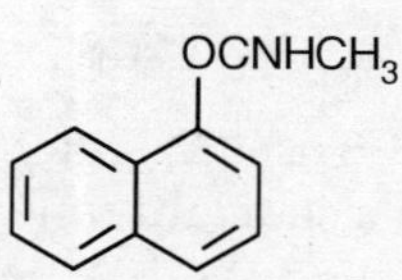

Carbaryl

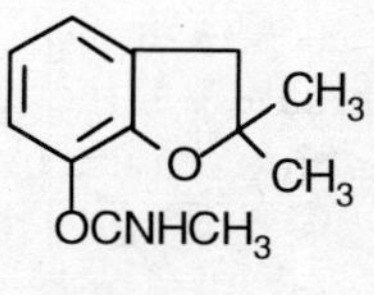

Carbofuran

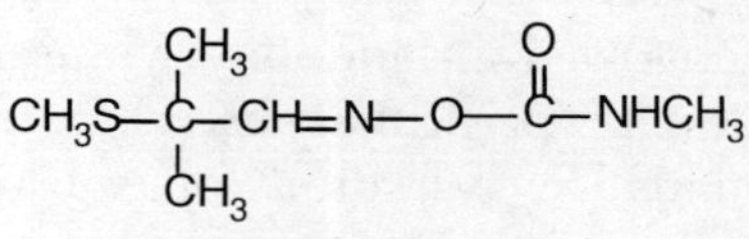

Aldicarb

## CHEMICAL MODIFICATION

Chemical modification of pesticides already of known structures are modified to develop better pesticides. For examples in 1941, Schradar prepared octamethyl pyrophosphoramide known as schradan with the following structure

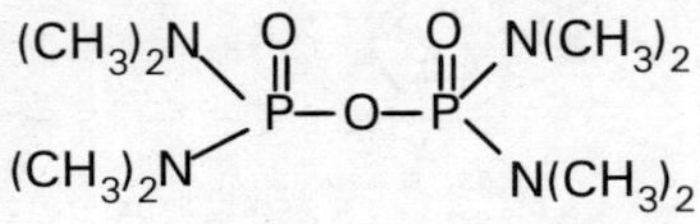

Octamethyl pyrophosphoramide

Based on the structure of schradan, teterethyl pyrophosphate and sulfotep were synthesed

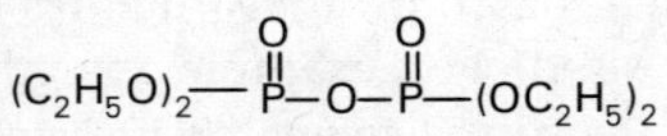

Tetraethyl pyrophosphate

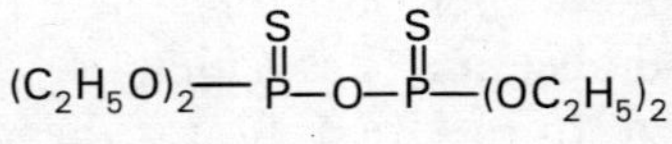

Sulfotep

## MATHEMATICAL MODELS

The bioactivity of pesticides depends on several processes taking place in the biological object, including permeation and transport, detoxication and interaction with the biological receptor. The properties are influenced by the different hydrophobic, electronic and steric properties of the compound. In a congeric series, these possible influences can be studied in a quantitative way by means of multiple regression analysis using substituent constants for the properties mentioned. This study known as "quantitative structure activity relationship" (QSAR) was initiated by C. Hansch in 1962 who had proposed the following mathematical models to correlate the biological activity with substituent constants.

$$\log 1 / CS = -a\pi^2 + b\pi + C\sigma + dEs + e$$

where CS is the concentration of member S which gives a standard response in a standard time interval (e.g. $ED_{50}$) a, b, c, d and e are constants and $\pi$, $\sigma$, and Es are hydrophobic, electronic and steric substituent constants, respectively.

The quantitative structure activity relationship is thus a rational approach for discovering new pesticides and can save a lot of time and money involved in randomly synthesising and evaluating large number of pesticides. It is particularly useful for optimisation of bioactivity in a class of compounds by variation of substituent patterns. In this approach, biological response is correlated with various hydrophobic, electronic and steric parameters of the members of series. This results in mathematic equations of several combinations of these parameters are also tried to obtain statistical equations from which the activities of not yet synthesised compounds of the same class can be predicted ultimately leading to compounds with the optimum activity.

The discovery and commercialisation of a new pesticide is a long term, multidisciplinary project. It has been estimated that it now requires the synthesis of 40,000 new compounds to produce one new product. The approximate cost of production of new pesticide is now put at over $ 60 million.

# CHAPTER 4

# Insecticides

Insecticides can be defined as any substance or mixture of substances intended for preventing, killing, repelling or controlling insects.

## CLASSIFICATION OF INSECTICIDES

Insecticides are classified on the basis of

1. Chemical nature
2. Mode of entry
3. Mode of action

### Based on Chemical Nature

Insecticides are classified based on the chemical nature as follows:

1. *Inorganic compounds:* arsenicals, fluorine compounds etc
2. *Organic compounds:* DDT, HCH, Carbaryl, phorate etc.
3. *Botanicals:* Nicotine sulphate, pyrethrum etc.

### Based on Mode of Entry

1. **Stomach Insecticides:** The insecticides are applied to that part of the crop which serves as food for the pest and is required in the normal process of eating, for example DDT, HCH. Their application is through food and entry through the midgut of the insect. Stomach poisons are necessarily, therefore to be ingested during feeding.
2. **Contact Poisons:** Contact poisons are those which kill the pest only due to contact or absorbed action. Their application is through the body surface and entry through the cuticle and tracheae. For example pyrethrum, rotenone, toxaphene, and chlorodane.
3. **Fumigants:** Gaseous poisons used for killing insects are called fumigants. These are applied in vapour state and their entry is through the tracheae e.g. HCN, chloropicrin, methyl bromide, DDVP, aluminum phosphide.
4. **Systemic Insecticides:** Systemic Insecticides are toxicants which when applied to the root, stem, or leaves of plants are rapidly absorbed and translocated to various parts of the plants in amounts lethal to insects feeding on these e.g. phorate, aldicarb, carbofuran. Translocation of these pesticides takes place mostly through xylem vessels only.

   The main properties of the systemic insecticides are:

a) The compound should have sufficient solubility to enable the compounds to move in the plant sap.
b) It should have ability to penetrate into the plant through roots, stems, and leaves.

Repper has given the following classification of systemics:

a) **Stable systemics:** Insecticides which don't metabolise in the system e.g. sodium fluorosilicate, sellinum.
b) **Endolytic systemics:** Systemic insecticides in which the toxic compound is present in 98% in its original form, which is taken up by the insects e.g. Schradan.
c) **Endometatoxic:** Systemic insecticides in which the toxic compounds gets converted into more toxic form, e.g. systox, metasystox.

### Based on Mode of Action

1. **Physical poisons:** These compounds exert a physical rather than a biochemical effect and these actually kill the insects by suffocation.
2. **Inert dusts:** They effect a loss of body moisture from insects by two types of action. Abrasive dusts e.g. aluminum oxide cause water loss by lacerating the epicuticle and water adsorbents e.g. charcoal remove water as a consequence of their hygroscopic properties.
3. **Heavy mineral oils:** These may exert a purely asphyrient effect killing scale insects slowly by exclusion of air.
4. **Protoplasmic poisons:** The action of these poisons apears to be associated primarily with the precipitation of proteins e.g. Heavy metals, nitrophenols, nitro cresols.
5. **Respiratory poisons:** They combine with cytochrome oxidase and other oxidases containing Fe and thus inhibit their catalytic action e.g. HCN, $H_2S$, CO.
6. **Nerve Poisons:** The action of these poisons is associated primarily with their solubility in the tissue lipids. e.g. chlorinated hydrocarbons, organophosphates, botanical insecticides.

## ORGANOCHLORINE INSECTICIDES

### DDT

(1,1,1,-trichloro,2,2-di (p-chlorophenyl) ethane)

This insecticide was first discovered by Zeidler (1874) but its powerful insecticidal properties were discovered by Paul Muller of the Swiss Geigy Company in 1939.

DDT is prepared by the condensation of chloral and chlorobenzene in the presence of excess concentrated sulphuric acid.

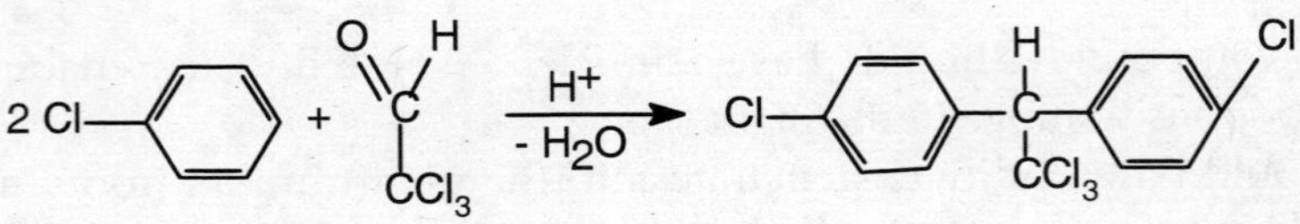

(1) 80%

+

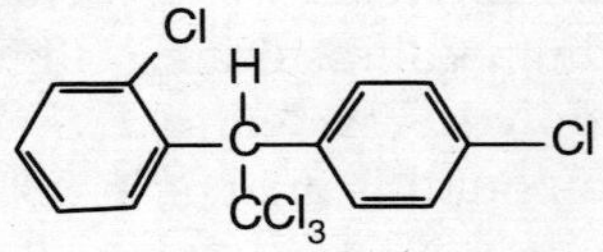

(2) 20%

The crude product consists of about 80% of the desired p,p' DDT (1) compound together with approximately 20% of the o,p' isomer (2) and trace of o,o' isomer. Only p,p' isomer has significant insecticidal activity. Pure p,p' DDT can be obtainted by recrystalization from ethanol.

***Reaction with Caustic Soda***

DDT on treatment with alcoholic KOH is converted into 1,1-dichloro-2,2-bis (p-chlorophenyl) ethylene (DDE) (3)

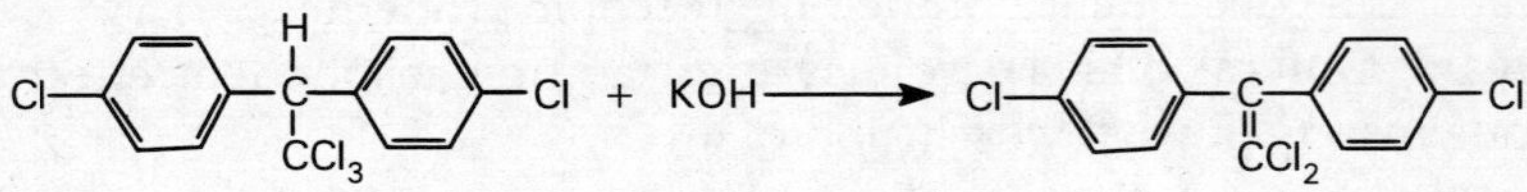

(3)

***Structure Activity Relationships***

1. DDT type molecule must be substituted in the p,p' position with halogen or short chain alkyl or alkoxy groups. Replacement of Cl by H, Br, I, OH, higher hydrocarbons, $NH_2$, $NO_2$, COOH, CN, CNS, groups decreases the activity than DDT. Replacement by F, $-OCH_3$- $CH_3$, $-C_2H_5$ does not decrease the insecticidal activity but mammalian toxicity is reduced.
2. Replacement of Cl of $CCl_3$ by F, Br, alkyl decreases the activity.
3. If H in CH-$CCl_3$ is replaced by OH, insecticidal activity is nil but the compound shows the acaricidal activity.
4. If $CCl_3$ is replaced by $CHCl_2$, $CH(NO_2)$, $CH_3$, $C(CH_3)_3$ the activity is decreased as follows.

$$CCl_3 > CHCl_2 > CH(NO_2) > CH_3 > C(CH_3)_3$$

5. If benzene rings of DDT are replaced by napthyl rings, then the compound is nontoxic to insects.

The metabolism of DDT is as follows

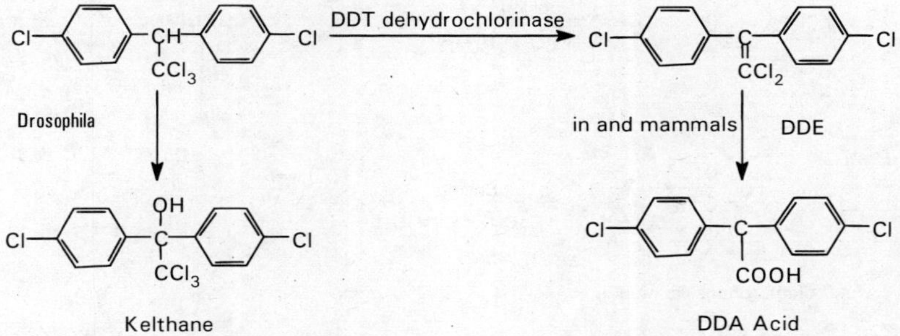


## Methoxychlor

(1,1,1 trichloro-2,2-bis (p-methoxy phenyl) ethene)

It is prepared in good yield by the condensation of anisole and chloral in the presence of sulphuric acid.

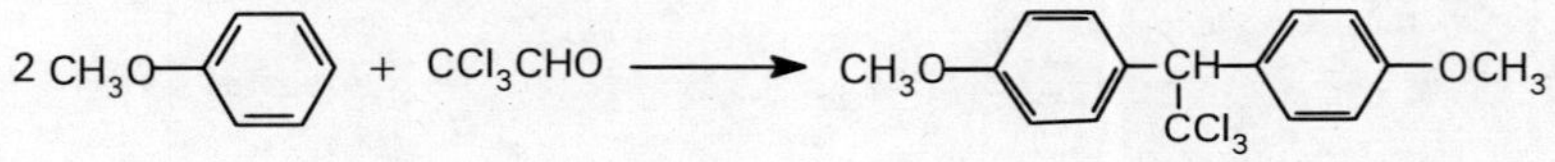

## DDD

(1,1-dichloro-2,2, bis (p-chlorophenyl) ethane)

Also known as TDE is prepared by the condensation of chlorobenzene with dichloroacetaldehyde

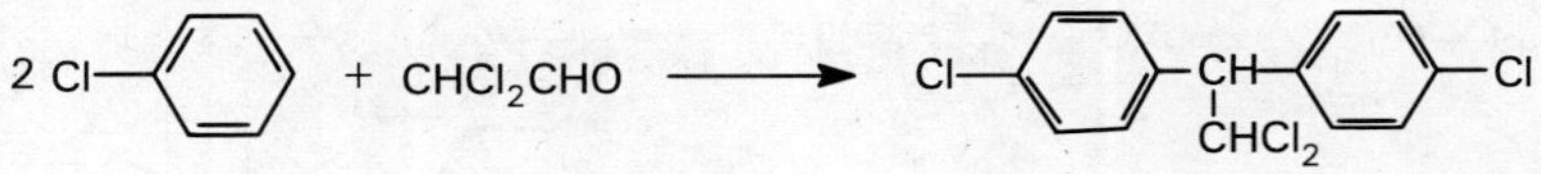

## Perthane

(1,1-dichloro-2,2 (bis (ethyl phenyl) ethane)

It is produced by the condensation of dichloroacetaldehyde with ethylbenzene in presence of concentrated sulphuric acid

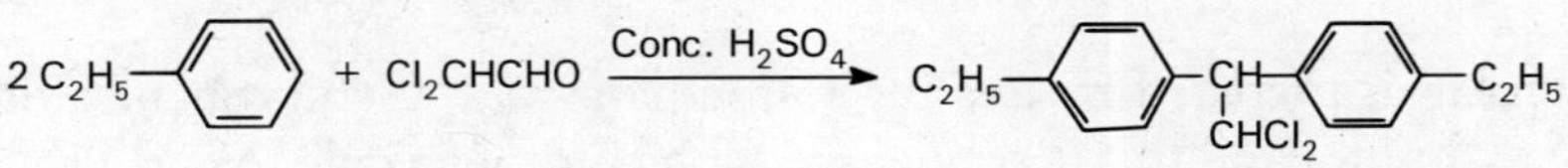

## DFDT

(1,1,1- trichloro,2,2-bis (fluorophenyl)

Ethane is prepared by treating fluorobenzene with chloral

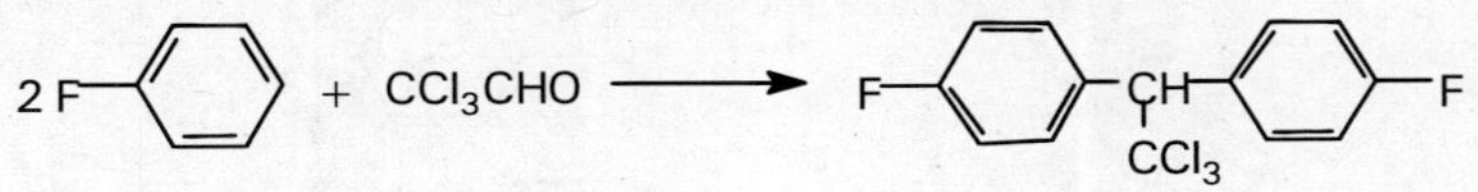

## Chlorobenzilate

(2-hydroxy 2,2- bis (p-chloro phenyl) ethyl acetate)

This is prepared from p-chlorobenzaldehyde via benzoin condensation.

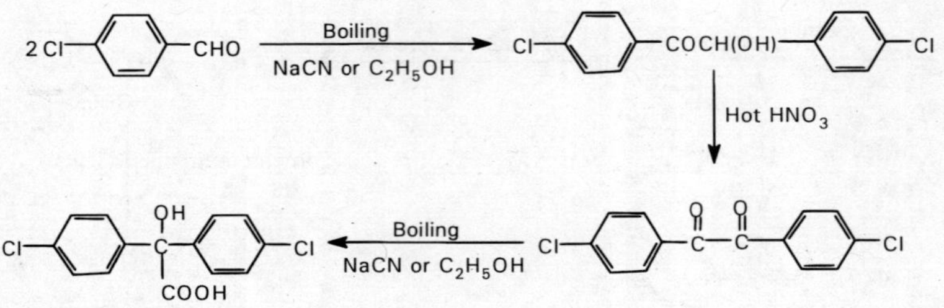

p,p'-dichlorobenzillic acid

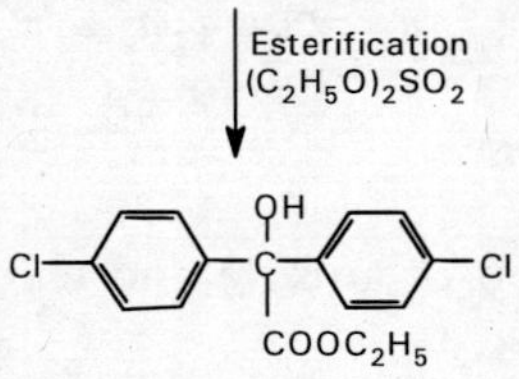

Chlorobenzillate

## Dilan

Dilan is a mixture of Bulan 1,1' (2-nitrobutylidene) bis -(4- chlorobenzene) and prolan (1,1-(2-nitropropylidene) bis (4- chlorobenzene). Bulan is prepared by addition of nitropropane to 4-chlorobenzaldehyde to give the carbinol which is then condensed with chlorobenzene to give Bulan

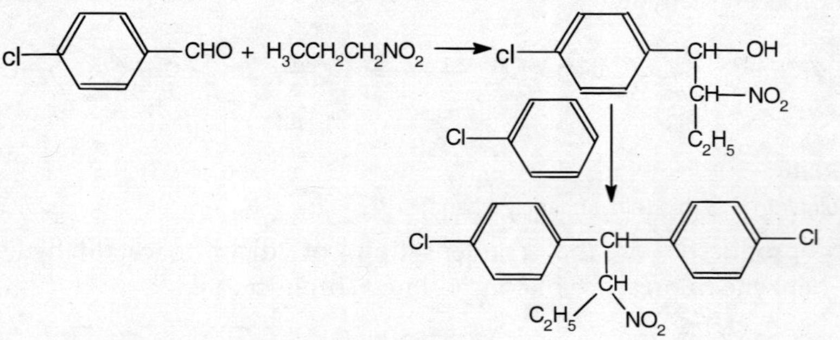

Prolan is prepared by treating nitroethane and 4- chlorobenz-aldehyde to give carbinol and this is condensed with chlorobenzene to give Prolan

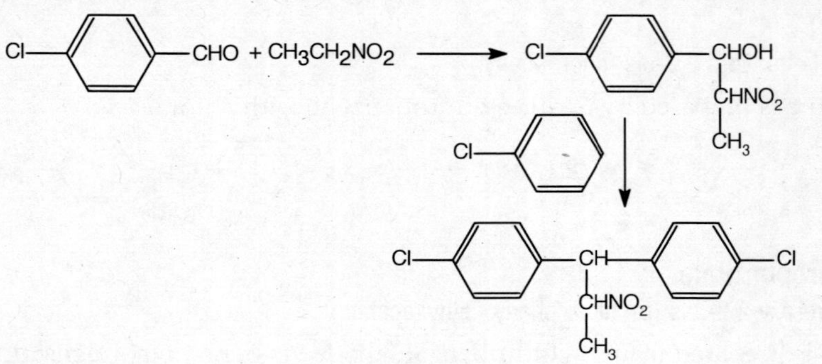

## HCH

(Hexacchlorocyclohexane)

HCH (Hexachlorocyclohexane) is prepared by teatment of benzene with chlorine under the influence of ultraviolet light without catalyst. Hexachlorocyclohexane can exist in εight different stereo-isomers of which five are present in product.

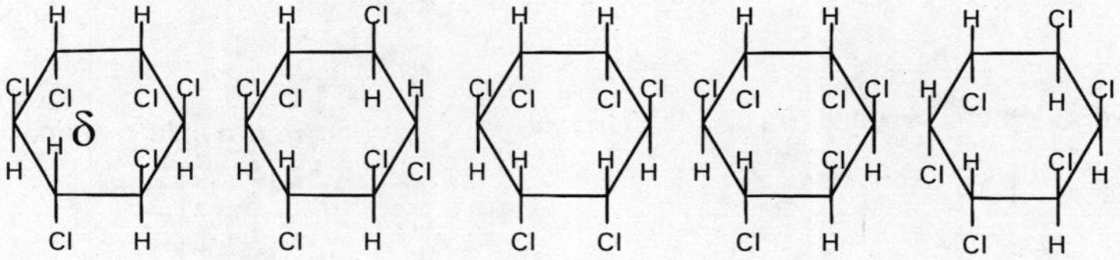

α ß γ δ ε

The chlorination is carried out in excess of methylene chloride. In chlorination of benzene, mixture of isomers of HCH is obtainted with the following contents.

- α-isomer 53.70%
- ß-isomer 3-14%
- γ-isomer 11-18%
- δ-isomer 6-10%
- ε-isomer 3-5%
- Heptachlorocyclohexane 3-4%
- Octachlorocyclohexane 0.5-1%
- Oily compounds 0.3-3%

The τ-isomer or lindane has insecticidal properties. The γ-isomer from the curde product is obtained by extraction with hot methanol followed by fractional crystallization to give pure γ- isomer. α and ß isomers are insoluable in methanol.

τ-HCH (Lindane) exerts fumigant action but not as DDT because it has an appreciably higher vapour pressure 9 x 10 $^{-6}$ mm Hg at 200°c than DDT (2x10$^{-7}$ mm Hg), so that it can exert a significant fumigant action.

### *Action with Caustic Soda*

With caustic alkalies in alcoholic solution, HCH splits into three molecules of HCl giving trichlorobenzene.

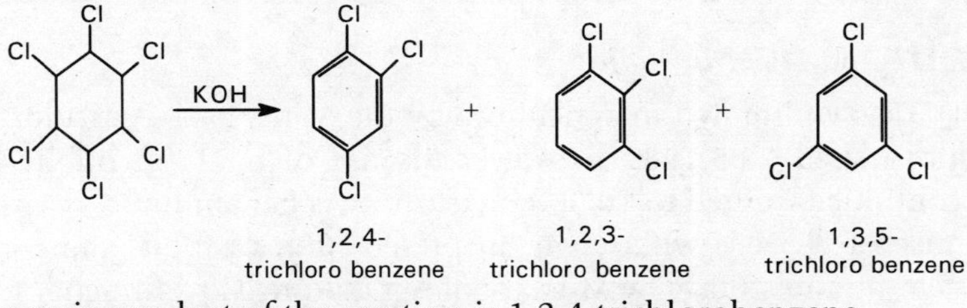


The main product of the reaction is 1,2,4-trichlorobenzene.

### *Metabolism of HCH*

HCH is metabolised in the plants as follows.

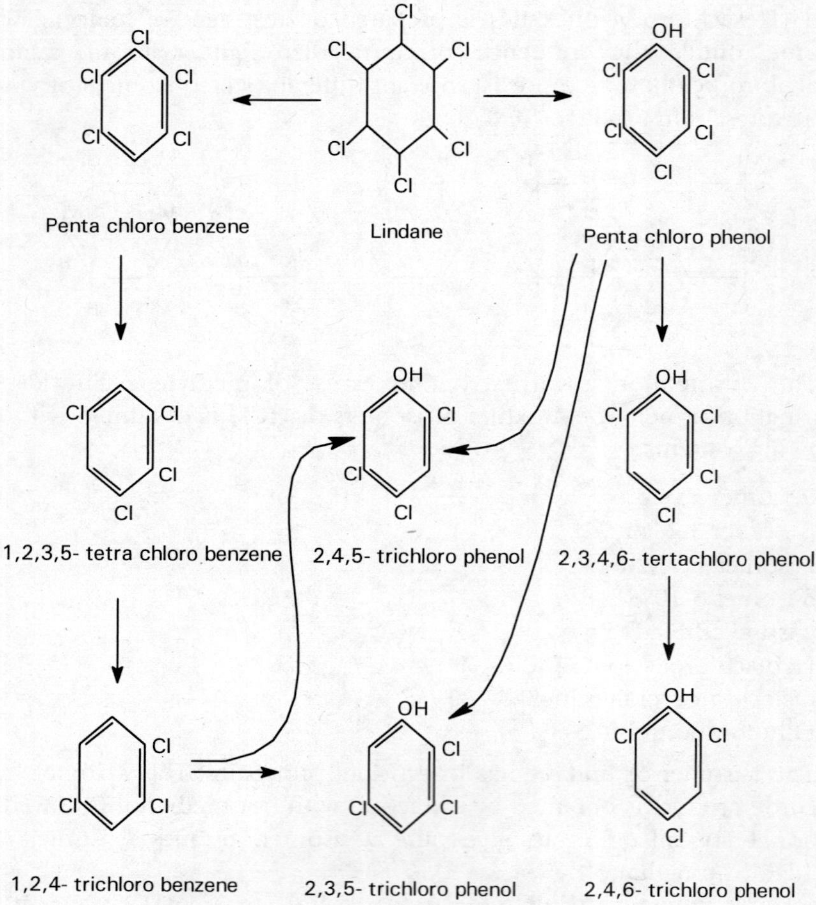


### *Structure Activity Relationship of HCH*

1. γ-isomer of HCH has insecticidal property and is about 50 to 10,000 times toxic than αand ß isomers.
2. If Cl in HCH is replaced by F, Br, I, $CH_3$, $OCH_3$, activity is reduced than γ-HCH
3. Six chlorine atoms are essential for activity.

### *Use of DDT and HCH in India*

Use of DDT is withdrawn in agriculture by the Ministry of Agriculture vide notification dated 26.5.1989. However the use of DDT for public health will be continued limited to 10,000 metric tonnes per annum.

The use of HCH is withdrawn on preservation of food grains, pulses, oilseeds etc. and in storage by Ministry of Agriculture, Government of India.

## THE CYCLODIENE GROUP INSECTICIDES

Most of the compounds of this group are prepared by Diel's-Alder reaction. A simple example of the Diel's-Alder reaction is the addition to conjugated diene of an ethylenic compound in which the double bond is adjacent to a carbonyl group e.g. butadiene combines with acraldehyde to form tetrahydrobenzaldehyde.

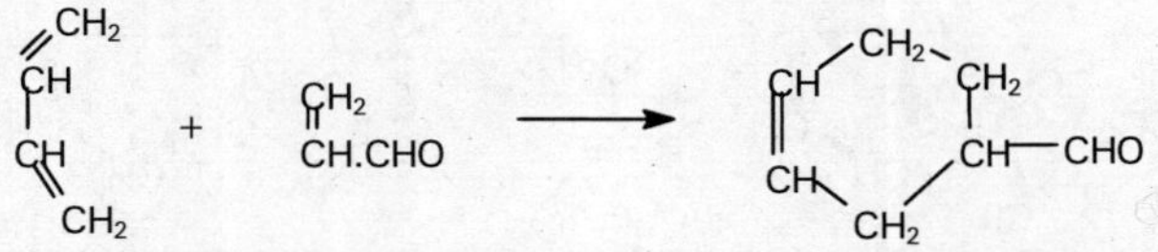

In general the diene synthesis is of the following type.

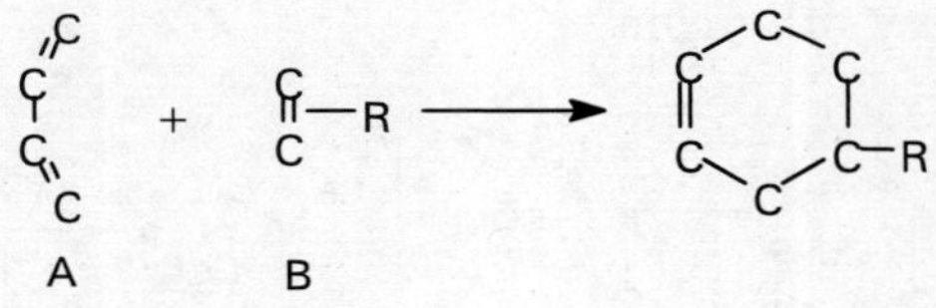

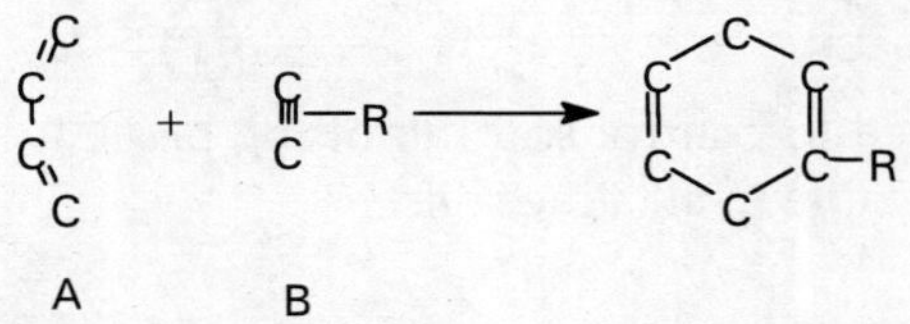

Compound A is refered to as diene and compound B is refered to as dienophile. R is usually a group which contains a carbonyl.

Characteristics of cyclodine Insticides:

1. They are polycyclic structures.
2. They are polychloro compounds
3. They contain methylene bridge
4. Their preparation is based on the Diels-Alder reaction.
5. Action of alkalies- Resistant to alkalies but chlorodane and toxaphene are hydrolysed.

### Heptachlor

(1,4,5,6,7,8,8, heptachloro-3a, 4,7,7a tetrahydro- 1,7-methanoindene)

Heptachlor is prepared by chlorination of chlordene with sulfuryl chloride. Chlordene is prepared by the reaction of hexachloro cyclopentadiene with cyclopentadine

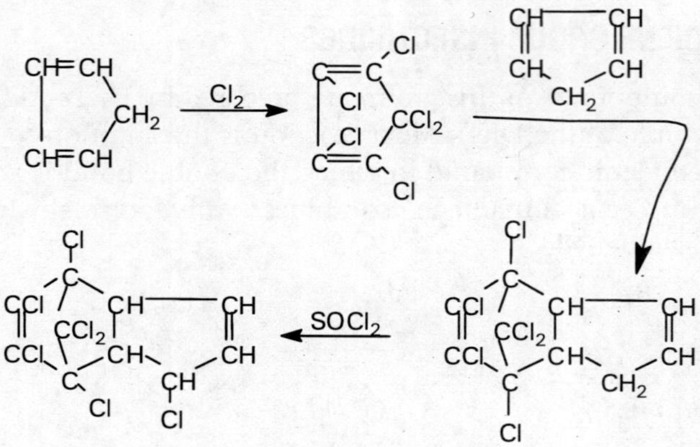

Heptachlor is also obtainted in good yield by the action of thionyl chloride with hydroxy chlordene.

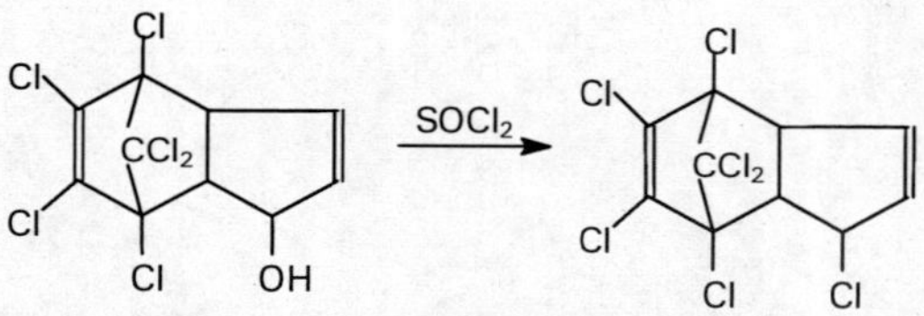

Heptachlor is used to control soil inhibiting pests. It has an advantage over BHC since it is stable in alkaline soils.

## Chlordane

1,2,4,5,6,7,8,8- Octachloro-2,3,3a,4,7,7a-hexahydro-4,7- methano-IH-indene

Chlordane is prepared by treating hexachlorocyclopentodiene with cyclopentadiene to give chlordene which is converted to chlordane by treating carbon tetrachloride or with sulfuryl chloride in the presence of iron chlorine

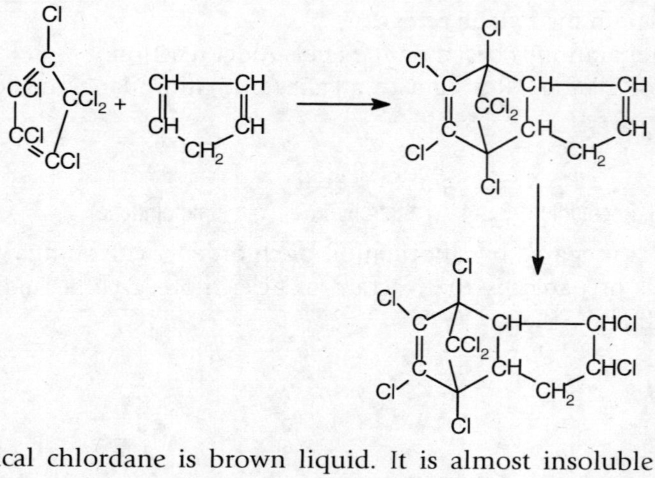

Technical chlordane is brown liquid. It is almost insoluble in water.

Chlordane is stable towards acid but losses hydrogen chloride under the influence of alkali. Its $LD_{50}$ is 283-590 mg/kg is broadly active contact, stomach and respiratory poison and is suitable for control of soil pests, white grubs and termites.

## Aldrin and Dieldrin

Aldrin ((1,2,3,4,10,10 hexachloro)-(1,4,4a, 5,8,8a hexahydro) -(1,4-endo, exo -5,8)- dimethano napthalene). Aldrin is obtainted by the reaction of hexachlorocyclopentadiene with an excess of bicyclo (2,2,1)-hepta-diene-2,5 at 100°C. Bicyclo ((2,2,1) -heptadiene-2,5 is obtained by the action of cyclopentadine and vinyl chloride followed by treatment with potassium hydroxide. Aldrin on epoxydation gives Dieldrin, 1,2,3,4,10,10-hexachloro,6,7 epoxy, 1,4,4a, 5,6,7,8,8a-octahydro-1,4-endo, exo-5,8-dimethanonapthalene

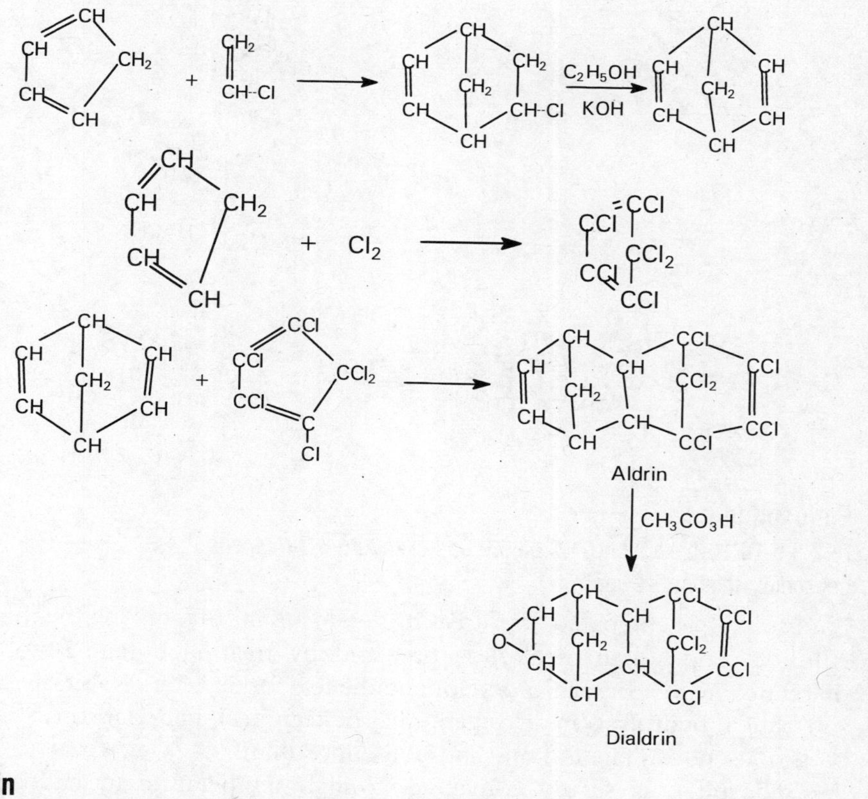


## Endrin

(1, 2, 3, 4, 10, 10 hexochlor-6, 7-epoxy-1, 4, 4a, 5, 6, 7, 8, 8a, octohydro, 14, 5, 8, endo-dimethano napthalene.)

Endrin is produced by the oxidation of isodrin with hydrogen peroxide in acetic acid.

Heptachlor, chlordane, aldrin, and endrin are used to control soil

inhibiting pests. These cyclodiene pesticides have an advantage over HCH since these are stable in alkaline soils.

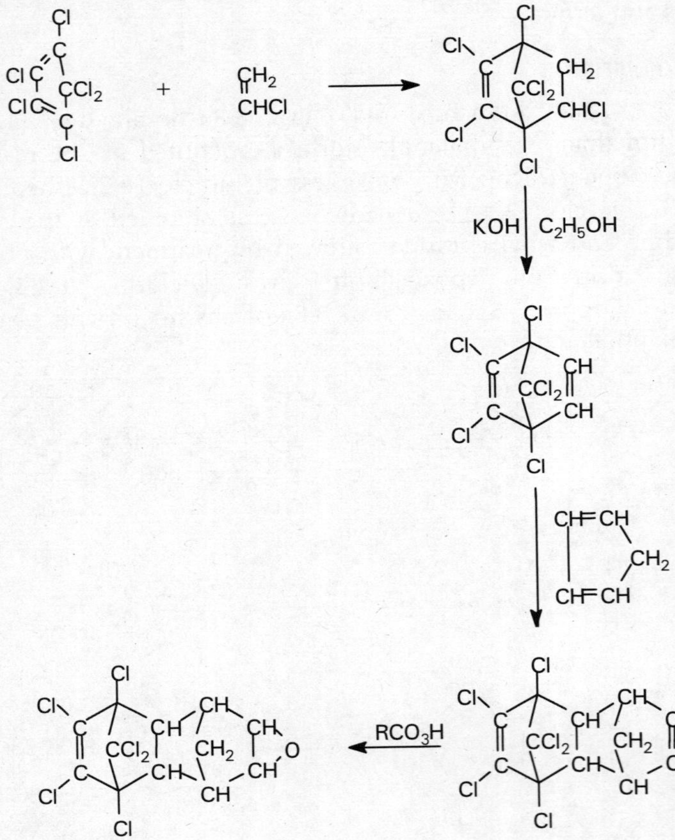


## Endosulfan

(6,7,8,9,10,10-hexachloro-1,5,5a,6,9,9a-hexahydro 6,9 metheno-2,4,3-benzodioxathiepin-3-oxide)

Endosulfan is prepared by treating endosulfan alcohol with thionyl chloride. Endosulfan alcohol is prepared by treating butene-2-diol-1,4-diacetate with hexachlorocyclopentadiene in xylene solution and subsequent hydrolysis in acid medium. The technical, material consists of 4 parts of α-endosulfan and one part of ß-endosulfan.

Endosulfan is slowly converted at high temperature to endosulfan sulphate. In plants, endosulfan is converted into different metabolites endosulfan sulfate, endosulfan ether and endosulfan alcohol.

Uses of endosulfan - Endosulfan is a broad spectrum non- systemic, contact and stomach insecticide and is used for control of pests on various crops.

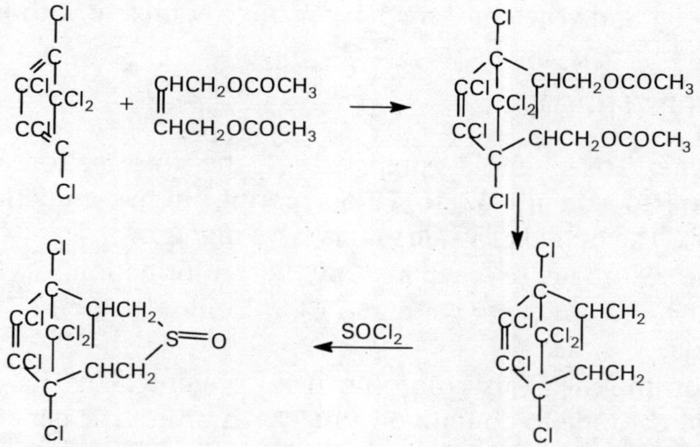

Endosulfan

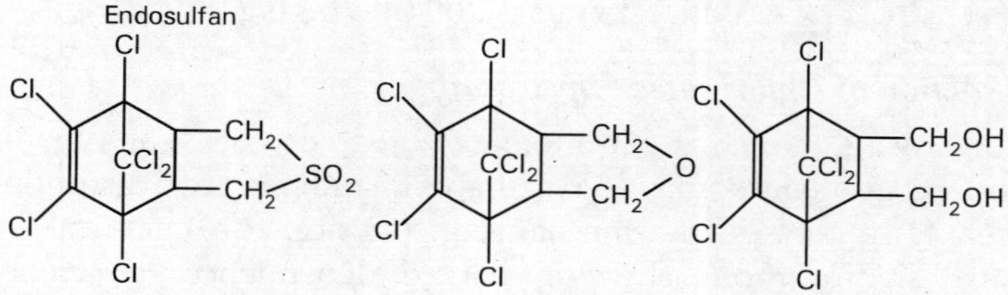

Endosulfan sulphate Endosulfan ether Endosulfan diol

### *Structure-Activity Relationship*

1. Increase in number of chlorine atoms in molecule increases the activity.
2. If all the chlorines are replaced by bromine, activity is reduced.
3. If two chlorine of cyclopentadine rings are replaced by F, then activity is increased.
4. Replacement of Cl of methano bridge by -$OCH_3$ group reduces the activity.
5. Highest activity is shown by these compounds that contain a polychlorinated centre together with another electronegative site such as double bound or Cl, O, S, or N atoms.
6. Configuration of rings, e.g. endo, oxo or endo, endo- has a substantial effect on insecticidal activity.
7. Removal of double bonds in aldrin by the addition of different groups like hydrogen, halogen, leads to decrease in activity.

### *Use of Cyclodiene Pesticides in India*

Registeration of endrin was withdrawn vide Ministry of Agriculture notification dated 15.5.90. Use of dieldrin is restricted for locust control by Plant Protection Adviser to Government of India only w.e.f. 15.5.90. Use of heptachlor and chlorodane has also been banned. Aldrin due to its proven termiticide properties and safety in low dosages is retained for termite control in soil and for seed treatment. But, its use is banned for use directly on food, fodder and vegetable crops or for directly mixing with food grains for preservation.

## CHLORINATED TERPENES

- **Toxaphene:** Toxaphene is prepared by the chlorination of bicyclic comphene to contain 67 to 70% chlorine. It has emperial formula $C_{10}H_{10}Cl_8$. The oral $LD_{50}$ value to rats is 69 mg/kg.
- **Storbane:** Storbane is prepared by the chlorination of mixture of comphene and pinene to contain 66% chlorine. The oral $LD_{50}$ to rats is 200 mg/kg.
- **Polychloropinene:** Polsychloropinene is obtainted by the chlorinatin of bornyl chloride to contain 64 to 67% chlorine. The general formula is $C_{10}H_{10}Cl_8$. The oral $LD_{50}$ to rats is 200-500 mg/kg.

### *Mode of Action of Chlorinated Hhydrocarbons*

Living organisms, which are elaborately ordered assays of organic and inorganic components, may be lethaly disordered by disrupting the essential life processes by mechanical, physical or chemical means. Chorinated hydrocarbons show a marked dependence of activity and structure and also cause specific modifications of the electrical properties of a nerve component.

### *Mode of Action of DDT and its Derivatives*

It is agreed now that the primary effect of DDT is on the nervous system in both, vertebrates and invertebrates as indicated by the following evidence.

1. Nervous impairment is indicated by DDT jitters along with hyperexcitability followed by ataxia (loss of motion and apparent paralysis.
2. When applied to isolated tissues and enzymes, only nervous tissues are sensitive to low concentrations.

A positive correlation between the level of DDT in central nervous system of a rat and the intensity of symptoms exist. (Dale et al, 1963).

All these evidences indicate that DDT is a neurotoxicant

Multiplication effect of DDT in nerve impulses was shown by Yeager and Munson (1945). Single nerve impulse arriving at a DDT trated region gives rise to prolonged volleys of impulses. DDT acts on axonic rather than on synaptic transmission.

Roeder and Weiant (1946) showed that sensory nerves were more sensitive than the motor nerves in case of American cockroach, which was later confirmed by Welsh and Gordon (1947).

Yamasoki and Narahashi, who started their study in 1957 have shown that in case of crural nerves of American cockroach, DDT increased the Negative After Potential' (NAP), which is associated with potassium efflux. They suggested that DDT inhibited this efflux. Matsumura and O'Brein

(1966) have finally shown that DDT greatly increased the K+ permeability of cockroach nervous tissues.

Brown (1951) reviewed the structure activity relationships in DDT analogues and concluded that any attempt to obtain any complete quantitative correlation with any single property ends in failure and that symmetry is the only hallmark of all the good DDT analogues.

Toxin Theory', introduced by Sternburg and Kearns (1952) suggests the presence of a blood borne paralysis factor in cockroach treated with DDT. This factor was also introduced by other chemicals like TEPP and dieldrin. This theory does not explain the primary lession, presumably the membrane destabilization.

A theory that DDT acts upon the cholinesterase also exists. Tobias et.al. (1946) reported a two fold elevation in acetylcholine level in the housefly or cochroach nervous system, at later stages of poisoning, while cholineterase was unaffected. Similar effects of lindane and aldrin have been reported but, Kearns (1957) found that much of the so called acetyl choline was, in fact, trichloroacetic acid from the extraction procedure. Colhoun (1960) concluded that the acetylcholine effect was only secondary effect of DDT poisoning. The effect of DDT on the acetylcholine synthesis was reported by Lewis (1953) in case of *Calliphora erythrocephata*. He showed that the synthesis was reduced first and then increased, indicating that some enzyme is first inhibited and then activated.

DDT has a negative temperature coefficient of activity, being more active at low temperatures in contrast to organophosphates, pyrethrins and several chlorinated hydrocarbons.

DDT causes a sharp and substantial increase in the oxygen consumption of insects (Harvey and Brown, 1951). This may be due to the increased muscular activity due to excessive nervous activity.

Disturbance of amino acid metabolism caused by DDT is also known. Carrigan and Kearns (1965) showed that treatment of American cockroach at 150° C caused decreas of 71% and 61% of tyrosine and proline, respectively and increase of 131% of phenylalanine with small changes in 11 other amino acids. Similar effect were found with DDT in housefly and with DDT in cockroaches.

In 1954, Millins proposed a theory that the activity of DDT is due to physically sleepe of its molecule, which is such that it fits into the intermolecular lattice. The insecticidal molecule entering the interspace of lipoprotein in the axonic memberane is firmly attached to the surrounding lipoprotein molecule. This brings some distortion which cause nerve 'disorder' of DDT molecule enters interspaces from its trichlorocarbon group. The dichlorophenyl carbon group spreads enough to make a triangle so that it constitutes two stable feet. In order to make this arrangement, the benzene ring should be able to rotate, DDE which posses the double bond between two carbon atoms is unable to fit in the interspace. Its double bound prevents the rotation of the benzene ring making it non-toxic. The structure-activity relationship of various isomers of DDT can be explained on the basis of this hypothesis. In p,p' DDT, the chlorine atoms are attached at para position which allows the rotation of phenyl group, but if chlorine atom is atached to ortho position as in o,p' DDT, the rotation of phenyl group becomes difficult and in o,o' DDT is becomes impossible. It is for this reason the insecticidal activity of various insomers is p,p'> m,p > m,o'> o,o DDT.

### *Mode of Action of Lindane and HCH*

Like DDT, lindane and BHC too are neuroactive agents as indicated by symptoms like tremors, ataxia convulsion and prostration, stimulated respiration, increase in acetylcholine levels, excessive electrical activity in poisoned axons (Lalonde & Brown 1954) and a negative temperature coefficient though smaller than that of DDT (Guthrie, 1950).

Comparative studies of individual isomers on various insects revealed that γ-isomer of BHC is 60-10000 times as active as, α, ß and δ isomers depending on insect species. Separation of γ isomer is important from view point of human toxicology, because on one hand it considerably decreases the total quantity of BHC used and on the other hand, α, ß and δ isomers which are without insecticidal effects, accumulates most easily in vaious organisms. The activity of γ isomer may be due to its molecular configuration. In gamma isomer, out of the six isomers of chlorine in benzene ring, three (1,2,3) are distributed alternately above and below at right angles to the equitorial plane of the benzene ring and rest of three atoms at 4,5,6 are located approximately in the equitorial plane of the benzene ring, no other isomers of BHC has this type of molecular configurtation.

According to Mullins (1955), the molecular diameter in the plane of ring of gamma isomer due to its special molecular configuration has a value of 8.5 A° for all the three maximum diameters where as other isomers have these values more than 8.5 A°. Because of this reason only γ-isomer can enter the interspace of lipoproteins causing distortion of pores and consequent excitation of nerve impulse transmission.

On the other hand lindane has been reported to be an antimetabolite of myornositol in a number of fungi, but it has been disproved in case of insects and mammals. Thus, BHC and lindane (γ-BHC) may be regarded as neurotoxicants.

### *Mode of Action of Cyclodiene*

Experiments conducted by Gianotti et al. (1956) with aldrin and dieldrin on the American cockroach have revealed that tremors in the legs were caused due to the effect of the poison on central nervous system. Effect of cyclodienes like chlordane, aldrin, dieldrin was shown by Lalonde and

Brown (1954) using a crural nerve of the American cockroach. Short trains of impulses, typically lasting 0.2 seconds were produced just as in the case of DDT.

Increased levels of acetylcholine in the nervous system as a result of aldrin and dieldrin poisoining have also been reported though the implications of this rise are obscure and one might guess it to be a consequence of nervous hyper-activity.

Studies by Hart et.al., Schultice et.al. and Fouts (1963) have shown that Chlordane have a stimulatory effct on hepatic cell microsomal activity for the metabolism of certain drugs, probably by stimulating the enzyme activity.

Orser and Brown (1951) reported that the injection of chlordane or toxaphene into Periplanata americana resulted in an irregularity of heart pulsation in both normal and decapitated insects. With chlordane the heart stopped in diastole and with toxaphene in systole.

***Mode of Action of Halo-Derivatives of Aliphatic Hydrocarbons***

Halogen derivatives of aliphatic hydrocarbons are biologically active and are characterised by their low vapour pressures, thus high volatility. They are thus fumigants and are considered to be physical poisons, killing by narcosis rather than biochemical reactions.

The narcotic action of these anaesthetics depends on the concentration achieved in some hypothetical lipid entity, called the biophase.

Shafer (1911) reported that they are rapidly absorbed into the insect, where they reduce the respiratory rate.

The dehydrogenase, which activates the formation of cytochrome, are inhibited by concentrations of narcotics.

## CARBAMATES

The backbone of all carbamate structures is carbamic acid HO-CO-$NH_2$ the monoamide of carbon dioxide. The acid does not occure in free form, since it decomposes into carbon dioxide and ammonia. Carbamic acid can be stabilized by the formation of simple alkyl ester such as ethyl carbamates. Most carbamates used as insecticides are esters of aromatic and hetetrocyclic hydroxy derivatives of N-methyl and N,N-dimethyl carbamic acids. These structural features are found in a natural product *Physostigma v-enenosa* which contain physostigmine (1) and prostigmine (2), the active compounds.

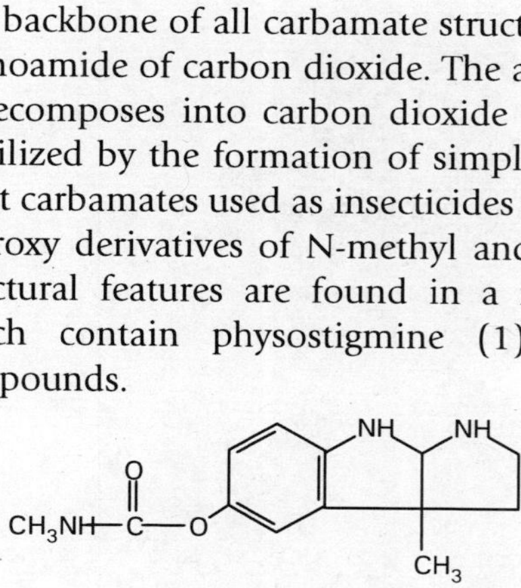

(1)

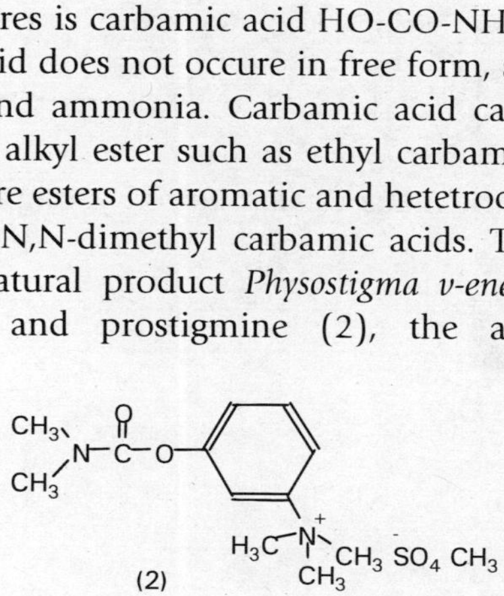

(2)

## Classification of Carbamates

Carbamates can be classified into following groups

1. Phenyl carbamates
2. N,N-dimethyl carbamates
3. Oxime carbamates

## PHENYL CARBAMATES

### Carbaryl

(1-Naphthyl-N-methyl carbamate)

Carbaryl is prepared by treating α- naphthol with methyl carbamoyl chloride or phosgene or methyl isocyanate.

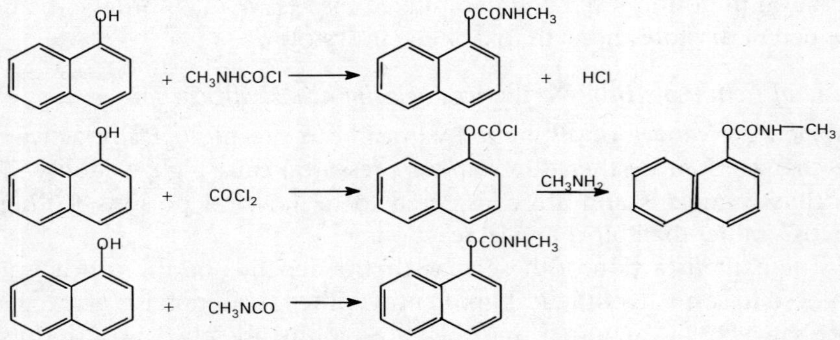


Carbaryl is a potent contact and respiratory poison and is very efficient against chewing pests. Carbaryl is a white crystalline compound m.p. 142° C. It is soluble to less than 0.1% in water but highly soluble in organic solvents. In alkaline medium, it is hydrolysed to α-napthol.

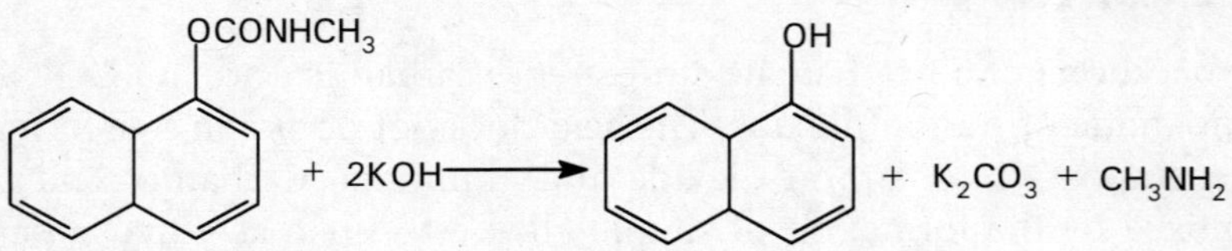


The metabolism of carbaryl on plants is shown below:

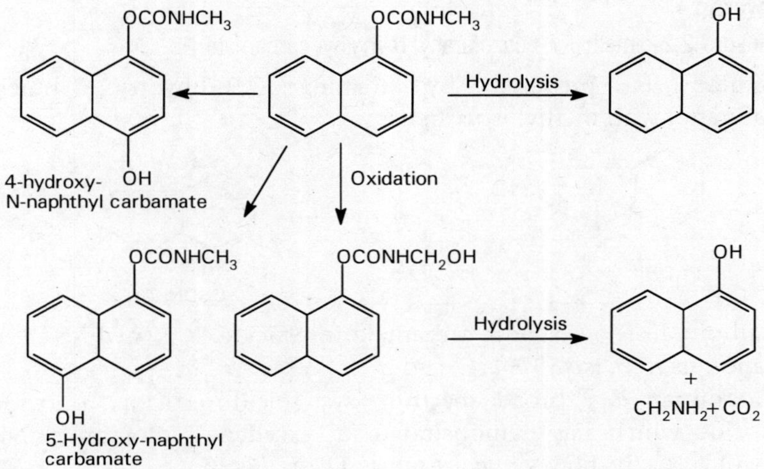


## Bendiocarb

(2,2-Dimethyl-1,3-benzodioxol -4-yl N- methyl carbamate)

Bendiocarb is synthesied by treating 2,2-dimethyl-1,3-benzodioxol-4-o1 with methyl isocyanate.

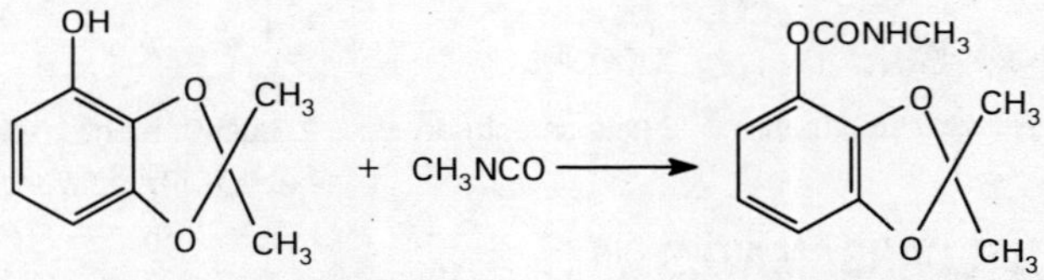


It is a wide spectrum insecticide with good residual activity following soil, foliar, surface application. The chemical has been commercially developed for the control of insects and other arthropod pests in and around buildings where it exhibits quick knowckdown properties.

## Baygon

(2-Isopropoxy phenyl-N-methyl carbamate)

Baygon is prepared by treating methyl isocyanate with 2- isopropoxyphenol

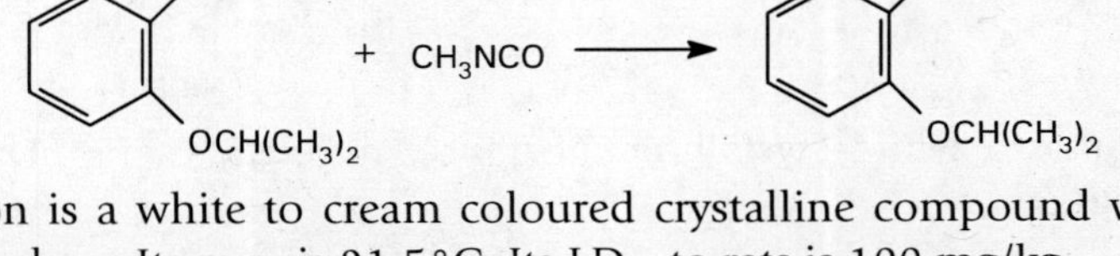


Baygon is a white to cream coloured crystalline compound with mild phenolic odour. Its m.p. is 91.5 °C. Its $LD_{50}$ to rats is 100 mg/kg.

Baygon has rapid knockdown properties and long residual action. It is particularly effective against insects affecting man and animals such as cockroaches, flies, mosquitoes. Its rapid knockdown and residual properties are important.

## Carbofuran

(2,3-dihydro-2,2-dimethyl-7-benzofuranyl N-methyl carbamate)

Carbofuran is prepared by treating 2,3-dihydro-2,2-dimethyl-7-benzofuranol with methyl isocyanate.

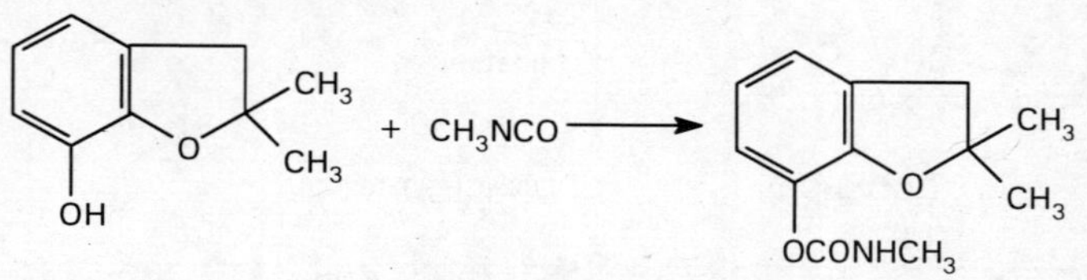


Carbofuran is a systemic carbamate insecticide. It is a white crystalline substance. Its m.p. is 153-54°C.

Carbofuran is a broad spectrum systemic insecticide, acaricide and nematicide which has demonstrated an excellent biological activity on variety of crops. Its $LD_{50}$ value for rats is 11 mg/kg

Important metabolites of carbofuran on plants are:

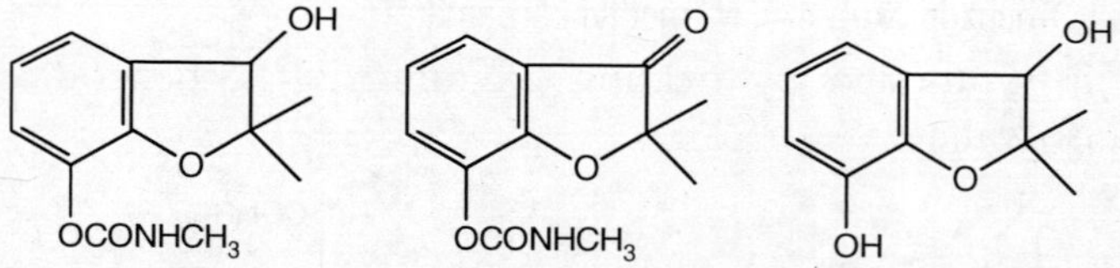


3-Hydroxy carbofuran    3-keto carbofuran    2,3-dihydro-7-hydroxy-2,2-dimethyl-3-oxobenzofuran

## N,N-DIMETHYLCARBAMATES

### Zectran

(4 (N-N-dimethylamino)-3,5-dimethylphenyl-N- methylcarbamate).

Zectran is prepared by treating 4-dimethylamino-3,5-dimethyl phenol with methyl isocyanate.

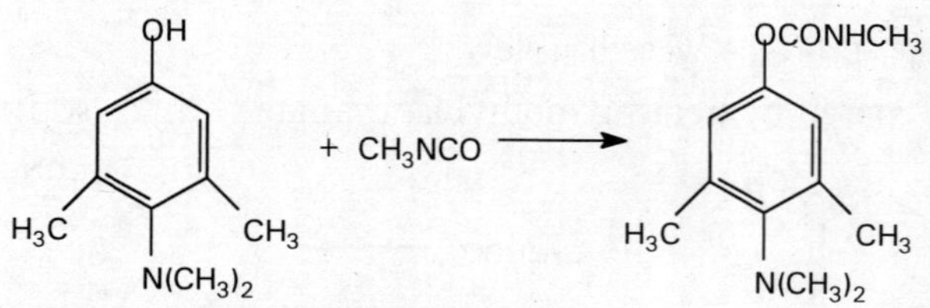


Zectran is a white crystalline substance, m.p. 85°C. Its $LD_{50}$ value is 160-63 mg/kg.

### Dimetan

(5,5-Dimethyl dihydroresorcinyl-N-N-dimethyl carbamate)

Dimetan is prepared by the reaction of monosodium salt of dimetan with dimethyl carbomoyl chloride.

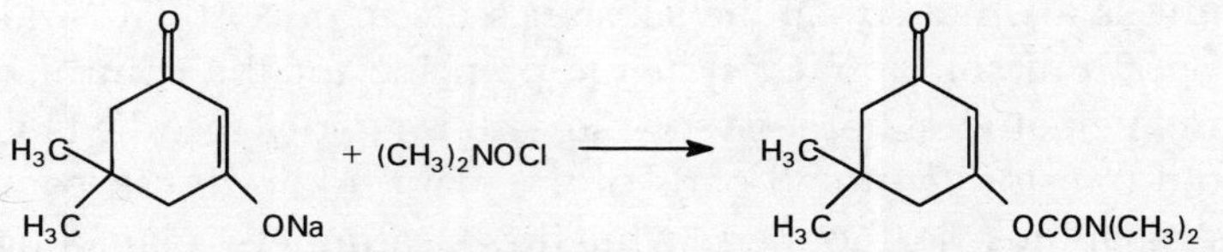

Dimetan is a yellow crystalline substance. The $LD_{50}$ for rats is about 150 mg/kg.

## Mesurol

(4-methyl mercapto-3,5-dimethyl phenyl N-methyl carbamate)

Mesurol is prepared by treating 4-methyl mercapto-3,5- dimethylphenol with methyl isocyanate.

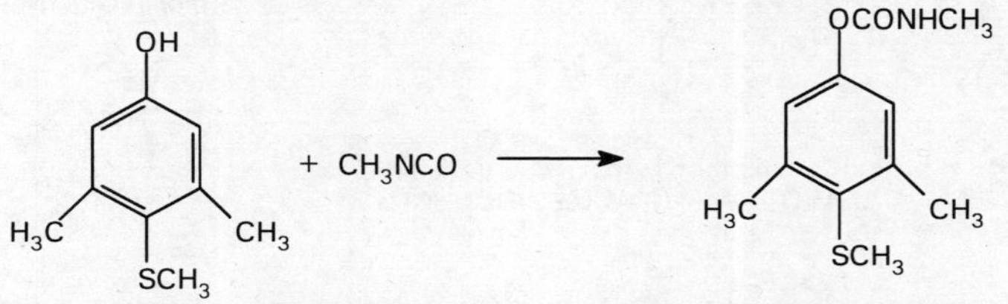

Mesurol is a white crystalline compound. Its m.p. is 121.5° C. Its $LD_{50}$ for rats is 100 mg/kg.

## Isolan

(1-Isopropyl-2-methyl pyrazolyl-5 dimethylcarbamate)

Isolan is prepared by the action of dimethyl carbamoyl chloride on the potassium derivative of enol form of 1-Isopropyl-3-methyl pyrazolone. Its $LD_{50}$ to rats is 11 to 23 mg/kg.

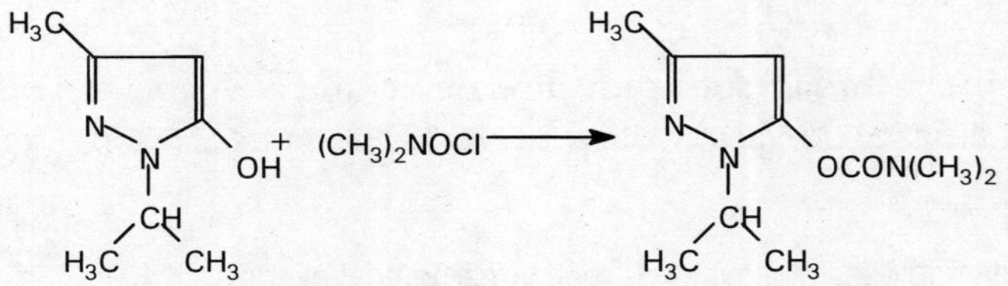

Isolan is a colourless liquid. Its b.p. is 105 to 107°C.

# OXIME CARBAMATES

## Temik

(Aldicarb, 2-methyl-2- methylthiopropional-oxime-o,N-methyl carbamates)

Aldicarb is obtained by the reaction of 2-methyl-2(methylthio) propionaldoxime with methyl isocyanate.

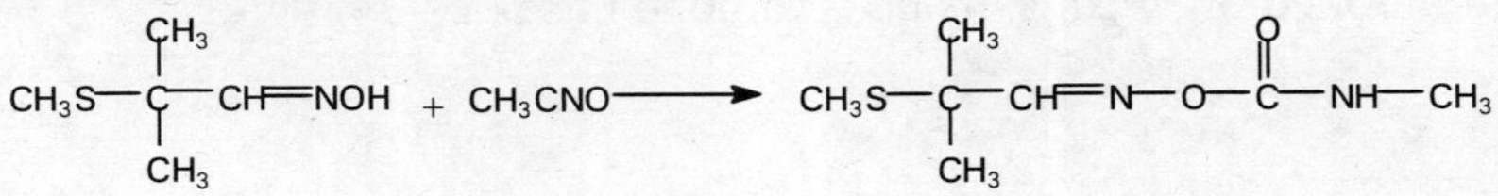

Temik is a white crystalline substance. Its m.p. is 100°C. Aldicarb is a broad spectrum soil applied systemic pesticide for the control of insects, mites and nematodes. In moist soil aldicarb is rapidly absorbed by the plant roots and translocated to all parts of the plant. Aldicarb can control many species of aphids, leafhoppers, whiteflies, leafminers, plant bugs, thirps, mealy bugs and mites. Its $LD_{50}$ value for rats is 0.93 mg/kg.

The metabolism of aldicarb on plants is shown below.

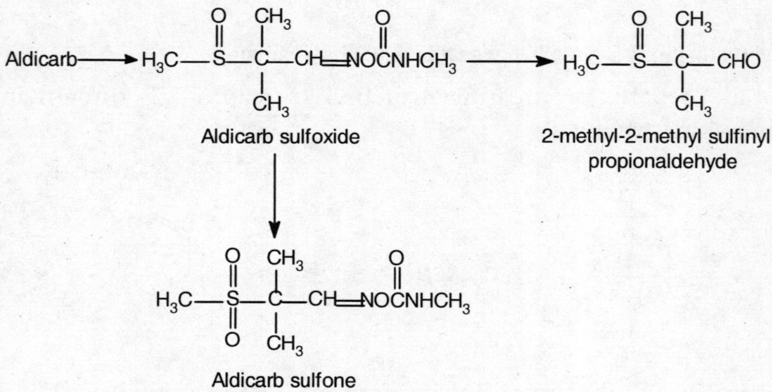


## Oxamil

S-methyl (1-dimethyl carbamoyl)-N (methyl carbamoyl) oxy) thioformomidate

Oxamil has unique systemic properties and is translocated both upwards and downwards in plants.

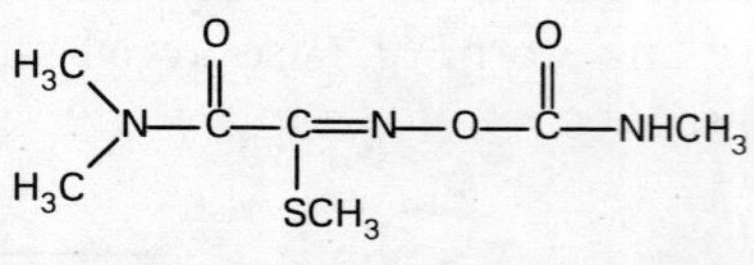

Oxamil is a broad spectrum insecticide and nematicide with excellent biological efficacy on variety of crops. Its $LD_{50}$ to rats is 5.4 mg/kg.

## Methomyl

(1-methyl mercaptoacetaldehyde- O- methyl-carbamoyl oxime)

Methomyl is a broad spectrum systemic insecticide and acaricide and is used for the control of cotton pests.

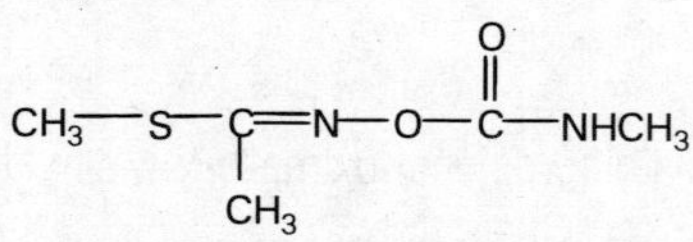

The spectrum of action is similar to aldicarb but it is highly toxic to warm blooded organisms. Its $LD_{50}$ value to rats is 17-24 mg/kg.

## Tirpate

(2,4-dimethyl-1,3-dithiobanecarboxaldehyde-O-methylcar bamoyl oxime)

Tirpate is prepared by treating mercaptol aldehyde with hydroxylamine and then carbamoylating with methyl isocyanate.

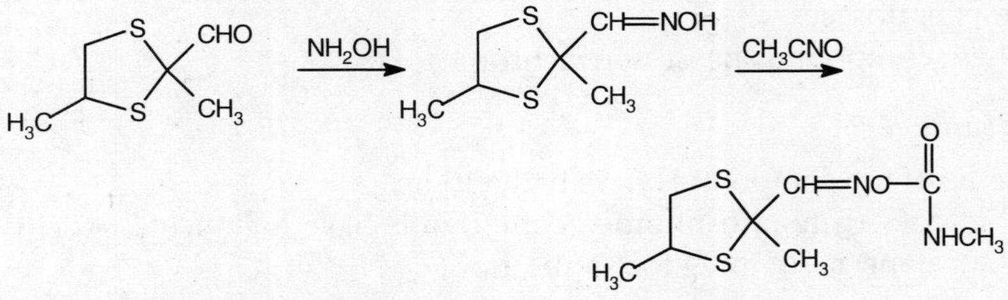


Tirpate is very toxic to warm blooded animals.

### *Mode of Action*

Mode of action of carbamates is described in the chapter of organophosphorous compounds under mode of action of organophosphorous insecticides.

### *Structure Activity Relationship of Carbamates*

1. Substitution of carbamate oxygen atoms by sulphur reduces toxicity considerably.
2. If the planer benzene ring is replaced by chair shaped cyclohexane ring, the activity is reduced.
3. Introduction of $CH_2$ group in between aromatic ring and carbamate group reduces toxicity.
4. If $CH_3$ group of $-OCONHCH_3$ replaced by other alkyl groups, then the toxicity is also reduced.

$$CH_3 > C_2H_5 > C_3H_7 > C_4H_9$$

5. Replacement of the second hydrogen atom on the nitrogen by an alkyl radical decreases the activity.

$$(CH_3)_3N > \text{tert-}C_4H_9 > (CH_3)_2N > \text{iso-}C_3H_7 > C_2H_5 > CH_3 > Cl > NO_2$$

6. Phenyl N-methyl carbamates with alkyl substitutent on the aromatic ring are most efficient when the alkyl group is in meta position with respect to carbamate group.

### *Organophosphorus Insecticides*

Organophosphorous insecticides form an important class of pesticides and about 100 insecticides are registered for use on various crops in the world.

Advantages of organophosphorous insecticides

1. Organophosphorous compounds are used as insecticides, acaricides, fungicides and herbicides.

2. They have broad spectrum of activity against number of pests.
3. They break down into non-toxic metabolites, hence there will be no pollution after their use.
4. Many of the organophosphorous compounds have systemic properties.
5. Due to high pesticidal toxicity, these are very economic and used in smaller dosages.
6. These compounds possess low chronic toxicity.

### *Disadvantages*

1. Some of the compounds have bad odours.
2. Some of organophosphorous compounds have high acute toxicity and require special training for application.

Different organic phosphorous compounds can be classified under the following heads. Derivatives of:

| | |
|---|---|
| 1. Phosphoric acid | 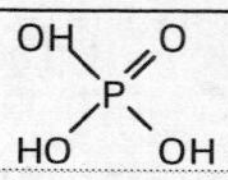 |
| 2. Thiophosphoric acid | 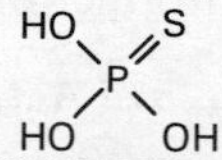 |
| 3. Dithiophosphoric acid | 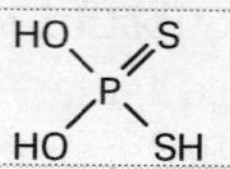 |
| 4. Phosphonic acid | |
| 5. Esteramides of phosphoric and phosphorothioic acid | |
| 6. Cyclic phosphates and phosphorothiolates | |
| 7. Pyrophosphoric acid | |
| 8. Cyclic phosphates | |

Reactions involved in the preparation of organophosphorous insecticides

### *Michaelis-Arbuzov Reaction*

This reaction is used for the preparation of phosphonate esters. In this reaction phosphonates are formed by the reaction of alkyl phosphites and alkyl halides. The intermediate is a quasiphosphonium salt which stablises by loss of alkyl halide and formation of phosphonate.

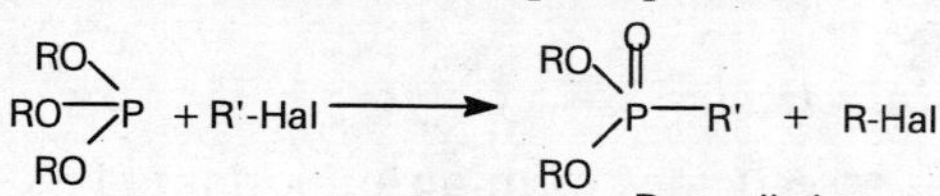

R = alkyl or aryl
R' = alkyl or aryl
Hal = Cl, Br, I

***Mechanism of the Reaction***

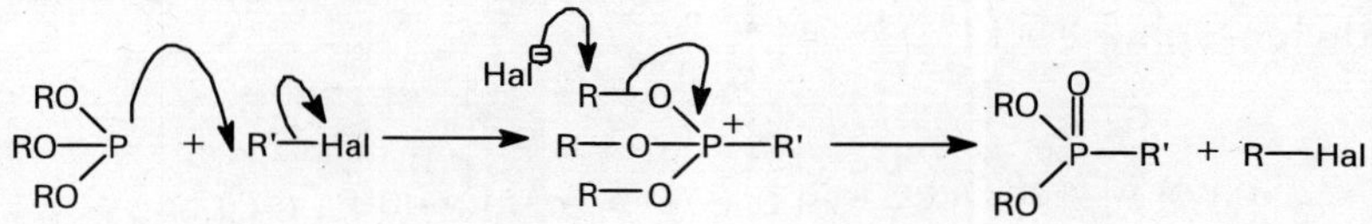

***Perkow Reaction***

In Perkow reaction, when trialkyl phosphites are treated with α-halocarbonyl compounds, vinyl phosphates are formed.

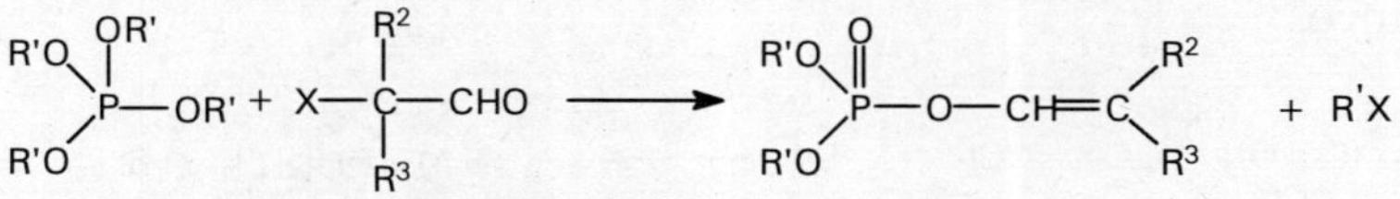

The proposed mechanisms is as follows

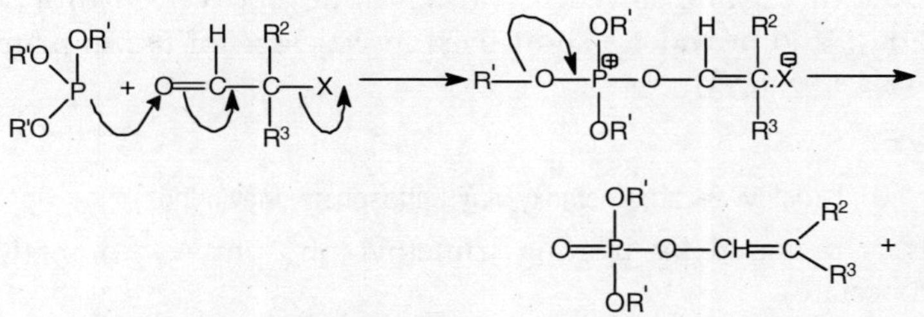

For example, dimethyl dichlorovinyl phosphate is produced from trimethyl phosphite and chloral

$$(CH_3O)P + O{=}CHCCl_3 \longrightarrow (CH_3O)_2\overset{O}{\overset{\|}{P}}{-}OCH{=}CCl_2 + CH_3Cl$$

## DERIVATIVES OF PHOSPHORIC ACID

### Dichlorovos

(O,O-dimethyl O-2,2-dichlorovinyl phosphate, DDVP, Nuvan, vaporoud)

DDVP is prepared by treating chloral with trimethyl phosphite

$$(CH_3)_3P + CCl_3CHO \longrightarrow CH_3Cl + (CH_3O)_2\overset{O}{\overset{\|}{P}}{-}OCH{=}CCl_2$$

It is also prepared by treating trichlorofon or diptrex with alkali in aqueous solution at 40-50 °C.

$$(CH_3O)_2\overset{O}{\overset{\|}{P}}CH(OH)CCl_3 + KOH \longrightarrow CH_3Cl + (CH_3O)_2\,P{-}OCH{=}CCl_2 + KCl + H_2O$$

DDVP is a colourless liquid, b.p. 120 °C at 14 mm Hg d, 1.425, nD25 1.4543. It is highly soluble in organic solvents. The $LD_{50}$ for rats is 80 mg/kg.

DDVP is hydrolysed in water and forms dichloroacetaldehyde and dimethylphosphoric acid.

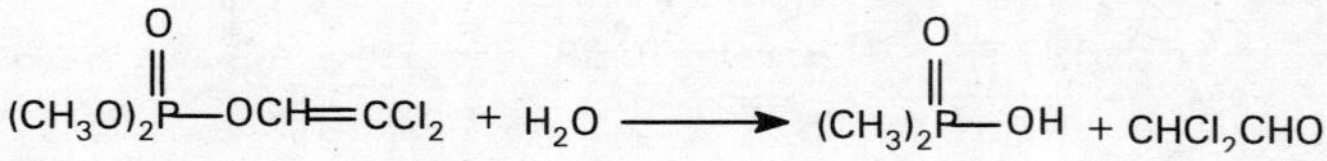

## Naled

(0,0-dimethyl-1,2-dibromo-2,2-dichloroethyl phosphate, Dibrom)

Dibrom is brominated DDVP and is prepared by brominaming dichlorovos.

It is a crystalline substance, m.p. 25°C, b.p.110°C. It is insoluble in water, soluble in organic solvents. Dibrom is destructively hydrolysed in water in 2 days and provides a short lived insecticides for use in plants. Its $LD_{50}$ for rats is 430 mg/kg.

## Mevinphos

(0,0-dimethyl 0-(1-methyl-2-carbomethoxy vinyl) phosphate, Mevinphos)

Mevinphos is prepared by treating trimethyl phosphite with methyl α-chloroacetoacetate)

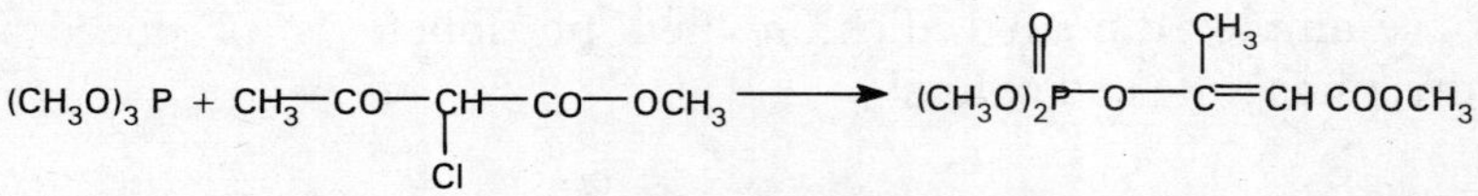

The commercial product is a mixture of cis and trans isomers containing 60% of the cis form and 40% trans. The cis isomer is 100 times more toxic to both insects and mammals than the trans form. It is a colourless liquid, highly soluble in water, b.p. 76°C at 0.2 mm Hg, vapour pressure at 20°C 2.2 x 10 mm of Hg, volatility at 26°C 27 mg/m$^3$. Its $LD_{50}$ to rats is 6.1 mg/kg.

## Phosphamidon

(0,0-dimethyl-0-(2-chloro-N,N-diethyl carbamoyl) methylvinyl phosphate, Dimecron)

Phosphamidon is prepared by the Arbuzov reaction accompanied by the Perkow rearrangement typical of α-halo carbonyls. Dichloroaceto N,N-diethyl acetamide is reacted with trimethyl phosphite to produce phosphamidon.

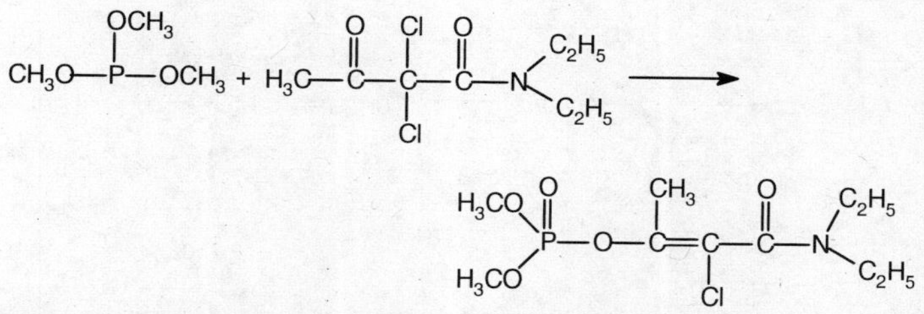

Phosphamidon, when pure, is a colourless, odourless oil. Its boiling point is 1600 C at 1.5 mm Hg vapour presure. It is soluble in water, alcohols, ketones and most highly polar oxygenated and halogenated solvents.

## Bidrin

(0,0-dimethyl- 0- (N-N-dimethyl carbamoyl-1-methyl vinyl phosphate)

Bidrin is prepared by Perkow reaction from dimethylamide of monochloroaceto acetic acid and trimethyl phosphite.

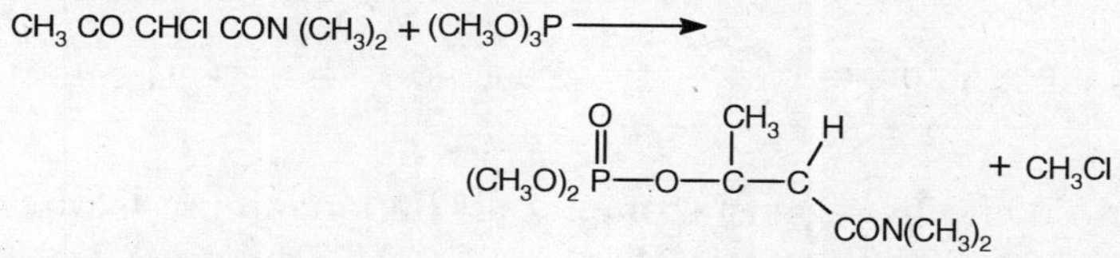

Bidrin is a brown liquid, b.p. 115-120°C at 0.001 mm Hg $d^{23}$ 1.19, soluble in water, ketones and alcohols. The $LD_{50}$ value to rats is 22 mg/kg.

## Birlane

(2-chloro-(2,4-dichlorophenyl) vinyl diethyl phosphate, Chlorfenvinphos)

Birlane is prepared by the condensation of triethylphosphite with 2,4,1',1'-tetrachloroacetophenone.

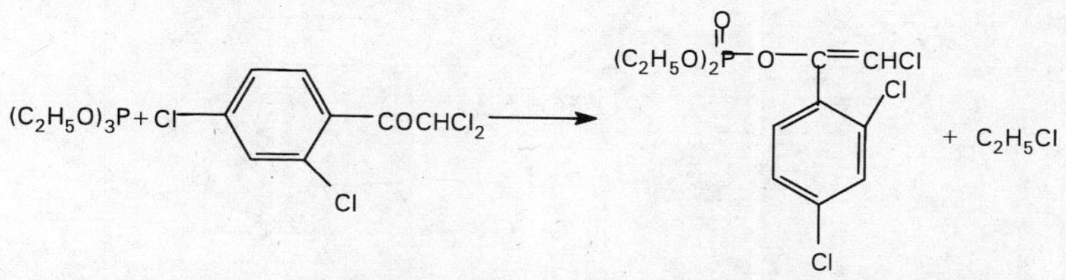

Birlane is an amber coloured liquid. The technical grade product is a mixture of cis and trans isomer in the ratio 8.5:1. The trans isomer is more active.

## Gardona

(2-chloro-1-(2,4,5-trichlorophenyl) vinyl dimethyl phosphate).

Gardona is prepared by condensation of trimethyl phosphite with 2,4,5,1',8- pentachloroacetophenone

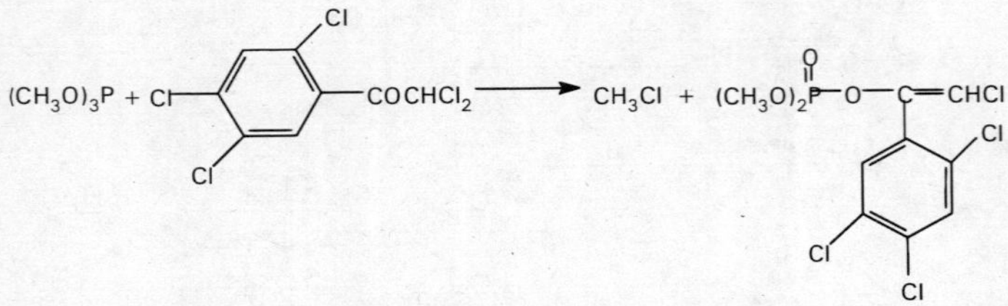

The crude product is a mixture of the geometric isomers; the trans isomer (with the chlorine atoms on oposite sides of the double bond) is obtainted (98%) by recrystallization. Gardona is a white crystalline substance, m.p. 97-98 °C. The $LD_{50}$ for rats is 4000 mg/kg.

## Azodrin

(Monocrotophos, Dimethyl phosphate of 3-hydroxy N-methyl cis-crotonamide)

Monocrotophos is prepared by treating timethyl phosphite with methyl amide of monochloroacetoacetic acid

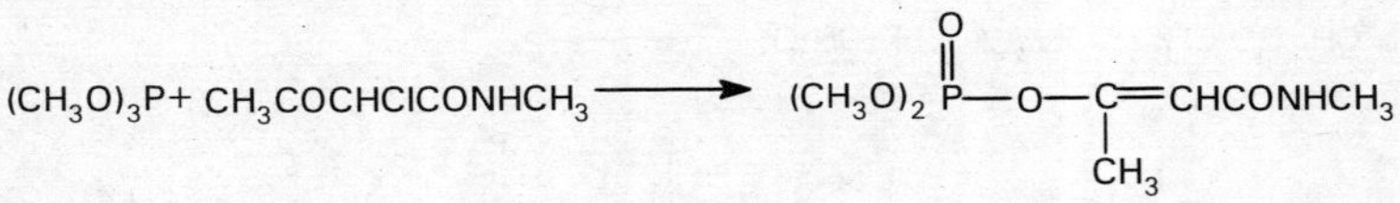

Azodrin is effective against economically important insects which attack on cotton. Pure azodrin is a solid, m.p. 53-55 °C. The acute oral $LD_{50}$ for rats is 21 mg/kg.

## Ruelene

(O-methyl O-(4- tert-butyl-2-chlorophenyl methylphosphoramidate)

Ruelene is prepared by reaction of 2-chloro-4-tertbutyl phenol with O-methyl-N-methyl amidochlorophosphate in the presence of base

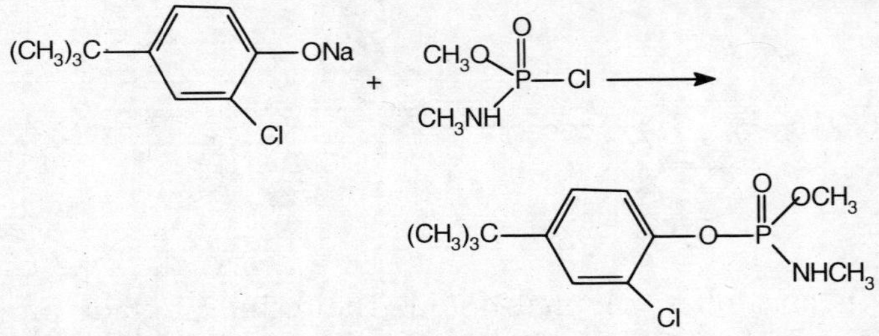

It is insoluble in water but soluble in organic solvents, m.p. 61 °C. The $LD_{50}$ for rats is 770-950 mg/kg. Ruelene is used for control of intestinal worms. It is also used for the control of ectoparasites of cattle.

## DERIVATIVES OF THIOPHOSPHORIC ACID

### Parathion

(O,O-Diethyl-O-4-nitrophenylthiophosphate, thiophos, E-605)

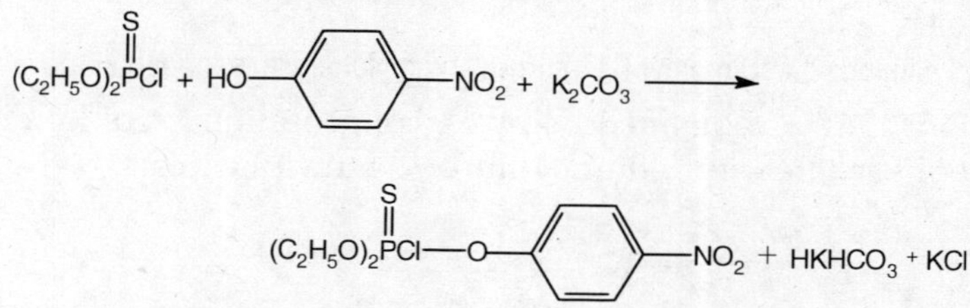

Parathion is prepared by reacting diethyl chlorothiophosphate with p-nitrophenol in the presence of potassium carbonate in acetone.

Parathion is a clear liquid, boiling point 157-162°C at 0.6 mm Hg solubility 24 ppm in water at 250C, miscible with aromatic solvents, lower molecular weight alcohols, acids, esters and chlorinated hydrocarbons. Its acute oral $LD_{50}$ to rats is approximately 4 mg/kg.

### Methyl Parathion

(Folidol, metaphos, O,O-Dimethyl-O-4-nitro phenyl thiosphophate)

Methyl parathion is prepared by reacting dimethyl chlorothiophosphate with sodium p-nitrophenolate in presence of potassium carbonate.

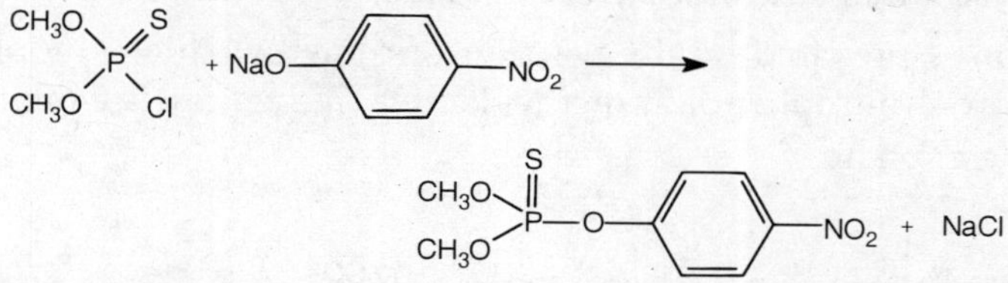

Methyl parathion is white crystalline solid with m.p. 35 to 36°C. It is used for the control of many insects and is effective against green leafhoppers, stem borers, and cutworms.

### Methyl Ethyl Parathion

(Thiophos, O-Methyl O-ethyl-4- nitrophenyl thiophosphate)

Methyl ethyl parathion is prepared by condensation of methyl ethyl chlorothiophosphate with p-nitrophenol and hydrochloric acid as acceptor in an organic solvent.

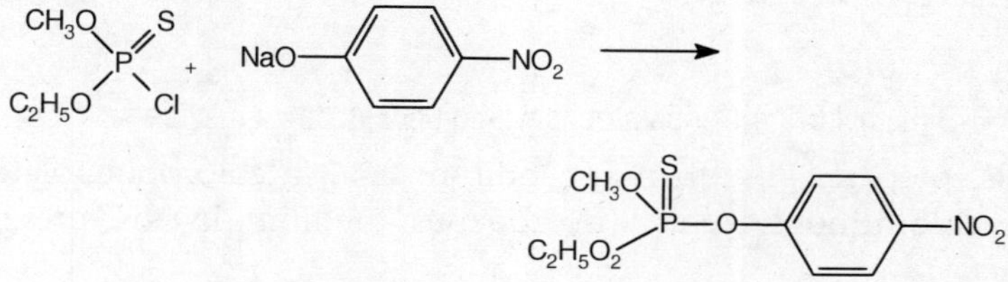

It is a clear liquid, b.p. 116°C at 0.12 mm of Hg vapour pressure at 20°C is 7.6 x 10-6 mm of Hg. Its solubility in water is about 40 mg/litre. It is soluble in most organic solvents except the paraffin solvents.

## Fenitrothion

(Sumithion, Folithion, 0,0-Dimethyl 0-4-nitro-3- methylphenylthiosphosphate).

Fenitrothion is obtainted by condensation of dimethyl chlorothiophosphate with sodium 4-nitro-3 methylphenolate

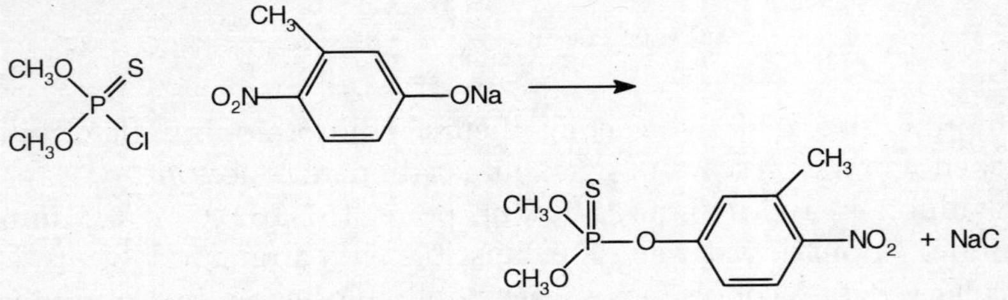

Fenitrothion is a liquid with unpleasant odour b.p. 95°C at 0.01 mm of Hg vapour pressure at 20°C. Fenitrothion is effective to control rice stem borers and is also useful to control flies and mosquitoes in public health programmes. The acute $LD_{50}$ to rats is 1000 mg/kg.

## Chlorothion

(0,0-Dimethyl 0-4-nitro-3-chlorophenyl thiophosphate)

Chlorothion is prepared by the reaction of dimethyl chlorothiophosphate with 3-chloro-4-nitrophenol in methyl ketone in the presence of anhydrous potassium carbonate.

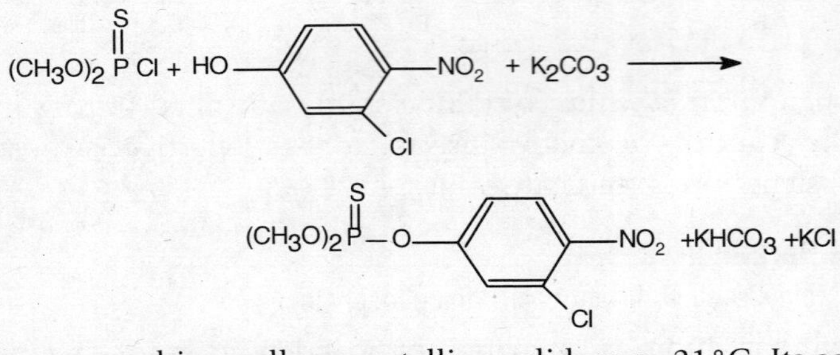

The compound is a yellow crystalline solid, m.p. 21°C. Its solubility in water at 20°C is 40 mg/litre. It is soluble in most organic solvents. The $LD_{50}$ for rats is 850 to 980 mg/kg.

## Ronnel

(Trolene, 0,0-Dimethyl 0-2,4,5-trichlorophenylthio phosphate)

Ronnel is prepared by treating sodium 2,4,5-trichlorophenolate with dimethyl chlorothiophosphate in aqueous medium in the presence of emulsifiers.

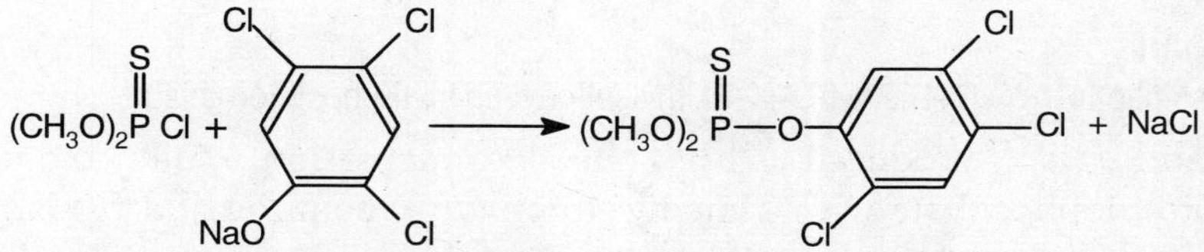

Ronnel is a white non hygroscopic powder with m.p. 36-37 °C, soluble in organic solvents. Its solubility in water is 0.01 ppm. The $LD_{50}$ for rats is 1250-1750 mg/kg. It has contact insecticidal activity against flies and cockroaches.

## Bromophos

(0,0-Dimethyl 0-2,5-dichloro-4-bromophenyl thio phosphate)

Bromophos is prepared by treating sodium salt of 2,5-dichloro-4 bromophenol with dimethyl chlorothiophosphate in aqueous medium in the presence of emulsifiers.

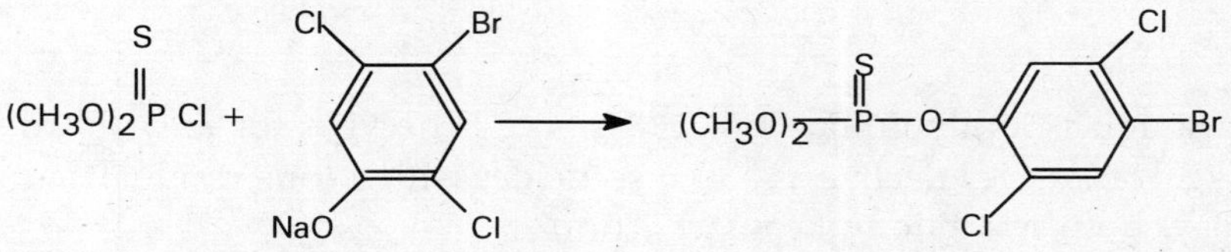

Bromophos is a white crystalline substance m.p. 54 °C, solubility in water at 20 °C is 40 mg/litre, and is soluble in organic solvents. The $LD_{50}$ for rats is 2000-4000 mg/kg. Bromophos is a nonsystemic insecticide and acaricide and is used on fruits and vegetables. It is useful as public health insecticide.

## Fenthion

(Lebaycid, 0,0-dimethyl-0-(4-methyl mercapto-3-methyl phenyl) thiosphosphate)

Fenthion is prepared by reaction of 4-methyl mercapto-3-methylphenol with dimethyl chlorothiophosphate in methyl ethylketone at 60 °C in the presence of potassium carbonate.

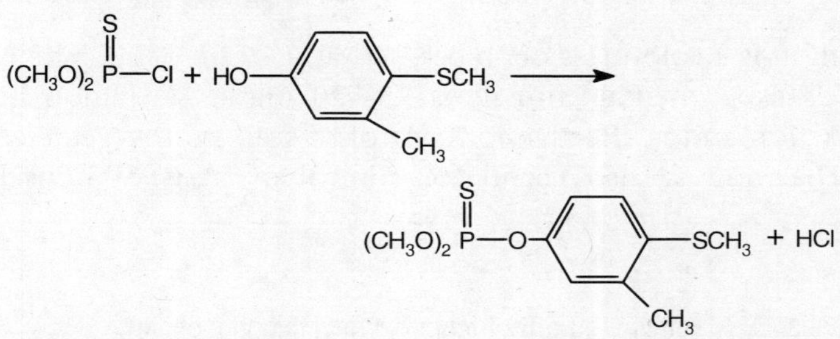

It is a colourless oil, b.p. 87 °C. It is effective for the control of mosquitoes, flies, ticks and cockroaches as well as pests on ornamental plants. Its acute oral $LD_{50}$ to rats is 215 mg/kg.

## Dasanit

(Fensulphothion, O,O -Diethyl-O-[4-methylsulfonyl phenyl] phosphorothiate)

Dasanit is produced by the condensation of o,o-diethyl chlorothiophosphate with 4-methyl thiohydroxybenzene and oxidation of the resulting ester with hydrogen peroxide.

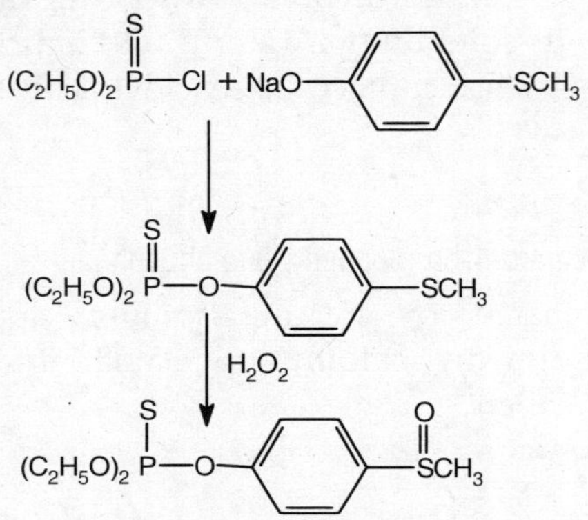


Dasanit is a oily liquid, b.p. 138-141 °C. The $LD_{50}$ for rats is 4 to 10.5 mg/kg. Dasanit is effective as soil insecticides with a long residual action. it displays both systemic and contact action.

## Diazinon

(O,O-Diethyl - O- (2-isopropyl-4-methylpyrimidyl-6) thio phosphate)

Diazinon is prepared by reacting 2-isopropyl -4-methyl-6 hydroxy pyrimidine with diethyl chlorothiophosphate in the presence of potassium carbonate.

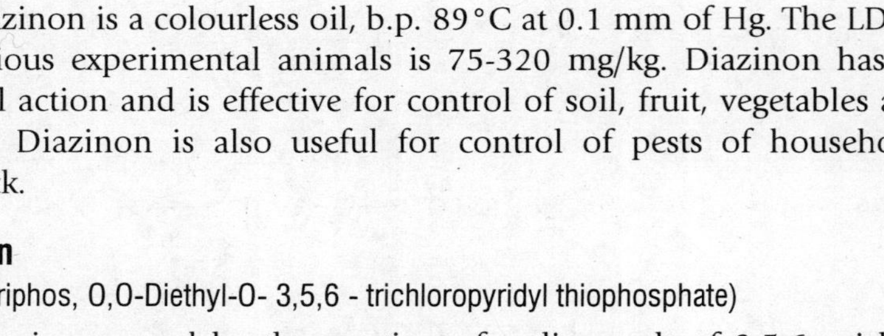


Diazinon is a colourless oil, b.p. 89 °C at 0.1 mm of Hg. The $LD_{50}$ value for various experimental animals is 75-320 mg/kg. Diazinon has a long residual action and is effective for control of soil, fruit, vegetables and rice insects. Diazinon is also useful for control of pests of household and livestock.

## Dursban

(Chlorpyriphos, O,O-Diethyl-O- 3,5,6 - trichloropyridyl thiophosphate)

Dursban is prepared by the reaction of sodium salt of 3,5,6- trichloro-2-hydroxypyridine with diethyl chlorothiophosphate in dimethyl formamide.

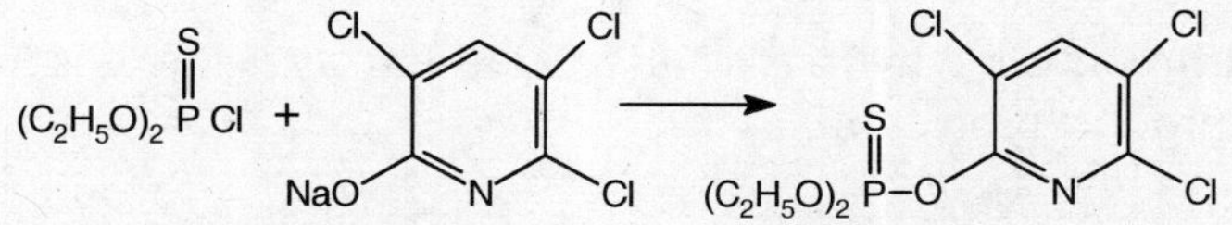

Dursban is a white crystalline solid, m.p. 41.5 to 43 °C. The $LD_{50}$ value for rats is 150 mg/kg and for guinea pigs 500 mg/kg. Dursban is used for control of sucking and chewing pests. It is also used for control of termites.

## Quinalphos

(Ekalux, O,O-diethyl-O-(2-quinoxaline) phosphorothioate)

Quinalphos is prepared by condesantion of O-phenylenediamine with hemiacetel of glyoxylate and then treating it with diethyl chlorothiophosphate.

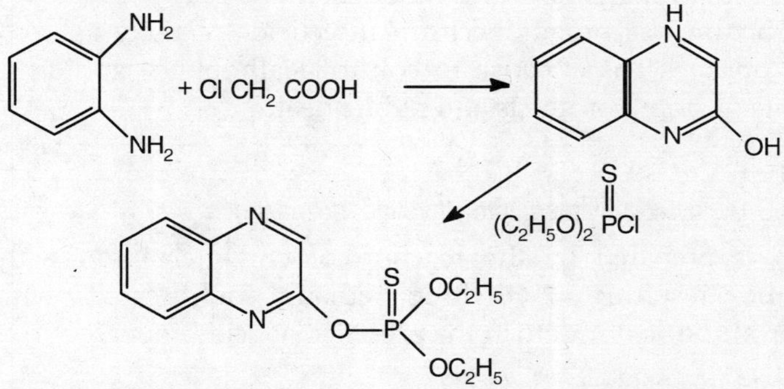

Quinalphos is a colourless and odourless crystalline solid with melting point of 31-32 °C. Quinalphos shows a high level of efficacy against a wide range of chewing, sucking and leaf miner pests. The oral $LD_{50}$ of the quinalphos for rats is 71 mg/kg and the dermal $LD_{50}$ 1750 mg/kg

## Pirimiphos Methyl

(O,O-dimethyl- O-(2-diethylamino -6- methyl-4-pyrimidinyl) phosphorothioic acid)

Pirimiphos methyl is prepared by condensation of ethylacetoacetate and N,N-diethyl guanidine.

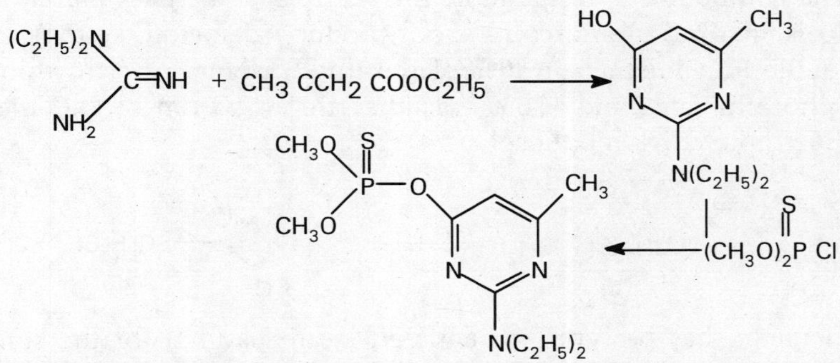

It is highly effective against sucking pests. Its $LD_{50}$ for rats is 2000

mg/kg. Pirimiphos methyl acts as insecticide and acaricide. It is mainly used in stored product protection.

## Phoxim

(O,O-diethyl phenylglyoxylonitrile phosphorothioate, Volaton)

Phoxim is prepared by treating α-cyanobenzaldoxime with diethylchlorothiophosphate

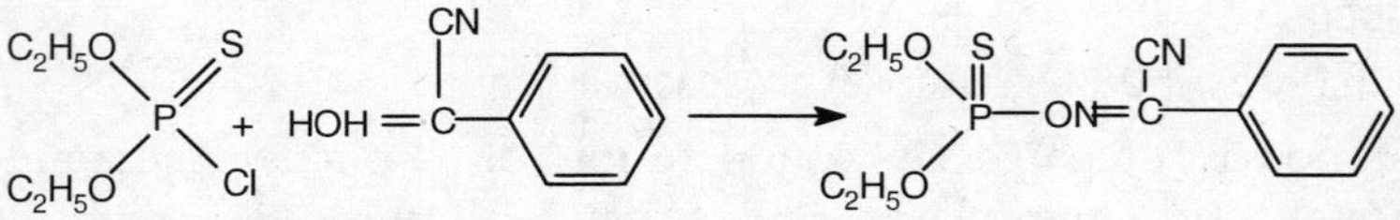

Phoxim is a light yellow liquid, b.p. 102°C at 0.01 mm Hg. Phoxim has low toxicity to human and warm blooded animals. The $LD_{50}$ for rats is 2000 mg/kg. Phoxim has broad spectrum insecticidal activity as a contact and stomach poison. It is effective to control leafhoppers, aphids, cut worms and insects of stored products and public health.

## Demeton

(Systox, O,O-Diethyl-2-ethylmercepto ethyl thiophosphate)

Demeton is prepared by the reaction of ethylmercapton with ethylene oxide, the resulting 2-ethylthio ethanol is treated with diethyl phosphorothionochloridate in the presence of acid binder.

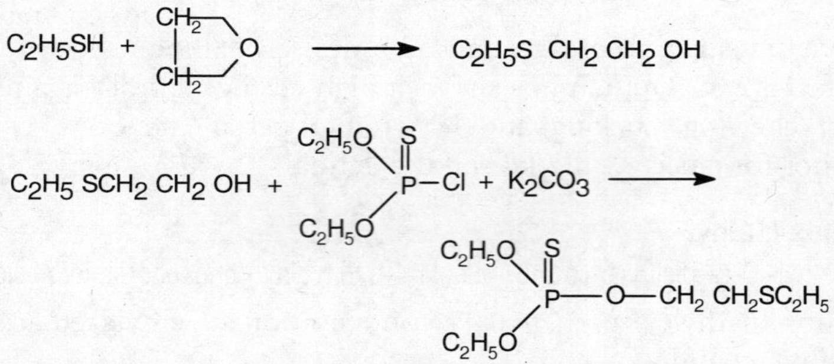

The compound is unstable in this form. It undergoes partial thiono-thiolo isomerization and commercial product (demeton, systox) contains an equilibrium mixture of 30% of the thiolo isomer of (ethylthio) ethyl phosphorothiolate and 70% of the thiono isomer O,O-diethyl-2-ethylmercapto ethyl thiophosphate.

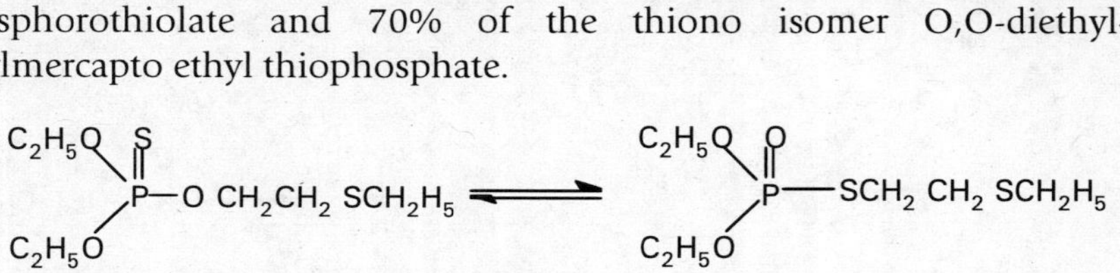

Demeton has powerful systemic effect and is used for the control of many sucking pests. The oral $LD_{50}$ value for rats is approximately 15 mg/kg.

In plant tissues both isomers are rapidly oxidised to the sulfoxide and sulfone derivatives.

## Triazophos

(O,O-diethyl O-(1-phenyl-1H-1,2,4 triazol-3-yl), phos phorothioate)

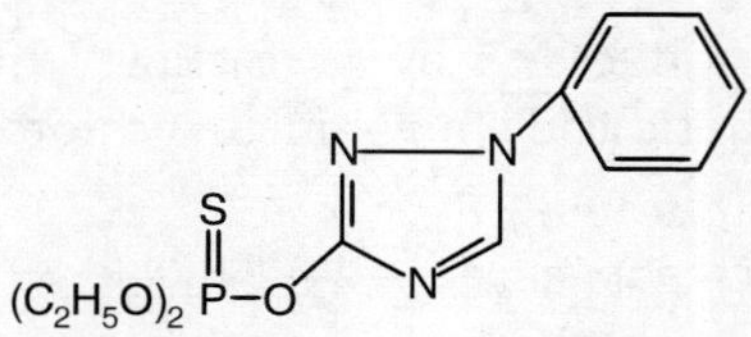

Triazophos is a broad spectrum insecticide effective against insects, mites, and nematodes. Its $LD_{50}$ to rats is 82 mg/kg.

## Methyldemeton

(Metasystox, O,O-dimethyl-2-ethyl mercaptoethyl thiophosphate)

Methyldemeton is the methyl homolog of demeton. Technical grade methyl demeton is a mixture of thiolo (30%) and thiono (70%) isomers,. Methyl demeton is prepared by the reaction of 2-hydroxy diethyl sulfide with dimethylchlorothiophosphate in the presence of HCl acceptor

$$(CH_3O)_2\overset{S}{\overset{\|}{P}}—Cl + HOCH_2CH_2SC_2H_5 + NaOH$$

$$\downarrow$$

$$(CH_3O)_2\overset{S}{\overset{\|}{P}}—O—CH_2—CH_2—S—C_2H_5 + NaCl + H_2O$$

Methyl demeton is less toxic than demeton. Thiono isomer has b.p. 93 °C and thiolo isomer b.p. 93 °C.

The $LD_{50}$ for thiono isomer of methyldemeton is 180 mg/kg and that of thiolo isomer is 40 mg/kg.

# DERIVATIVES OF DITHIOSPHOSPHORIC ACID

## Malathion

(O,O-dimethyl-S-(1,2-dicarbethoxyethyl-dithiophosphate)

Malathion is synthesised by reacting dimethyldithiophsphoric acid and maleic acid ester in the presence of basic catalysts.

$$(CH_3O)_2\overset{S}{\overset{\|}{P}}SH + \begin{matrix} CH\,COO\,C_2H_5 \\ \| \\ CH\,COO\,C_2H_5 \end{matrix} \longrightarrow (CH_3O)_2\overset{S}{\overset{\|}{P}}—S—\begin{matrix} CH\,COO\,C_2H_5 \\ | \\ CH\,COO\,C_2H_5 \end{matrix}$$

Malathion has a broad spectrum of action and is effective against both sucking and chewing insects on fruits and vegetables. It is also used for the control of flies and mosquitoes. Malathion is least toxic to warm blooded animals and its $LD_{50}$ values for rats is 1200 mg/kg. The metabolism of malathio in plants takes place as per the following scheme.

## Dimethoate

(Rogor, O,O-Dimethyl-S-(N-methyl-carbamoylmethyl) di- thiophosphate)

Dimethoate is produced by reacting potassium salt of dimethyldithiophosphoric acid with N-methyl chloroacetamide.

$$(CH_3O)_2\overset{S}{\overset{\|}{P}}-SK + Cl\,CH_2CONHCH_3 \longrightarrow (CH_3O)_2\overset{S}{\overset{\|}{P}}-S-CH_2\text{-}CONHCH_3 + KCl$$

Dimethoate is a systemic insecticide. It is used for control of insects on vegetables, cotton, tobacco and sorghum. Dimethoate decomposes slowly in plant tissues. Three processes occur simultaneously namely hydrolysis of the P-X bond, O-dimethylathion with the formation of products having low toxicity, and oxidation to the P=O analogue of dimethoate, a more powerful inhibitor of cholinestrase.

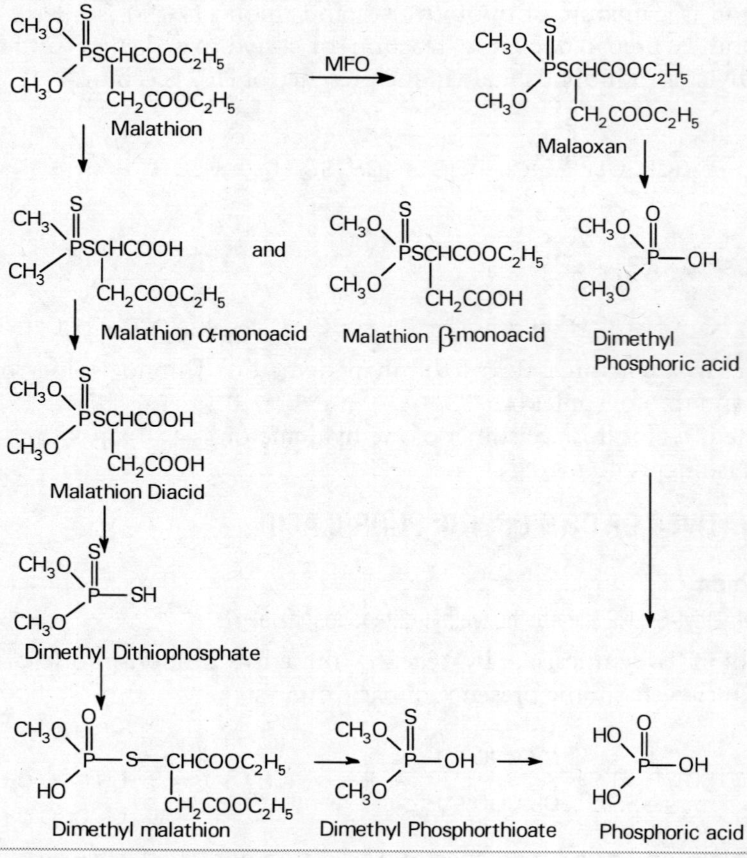


Metabolic pathway of malathion

## Thimet

(Phorate, O-O-Diethyl-S-(ethylthiomethyl) dithiophosphate)

Thimet is prepared by reacting diethyl dithiophosphoric acid with formaldehyde and ethyl mercaptan.

Phorate is systemic insecticide and is effective against sucking pests. Phorate is used as granules in soil and has relatively unstable to hydrolysis. However, it protects the plants for a relatively long time because of the greater persistency of the sulphoxide metabolite in plants and soil. Phorate is oxidised to sulphoxide and then to sulphone in plants.

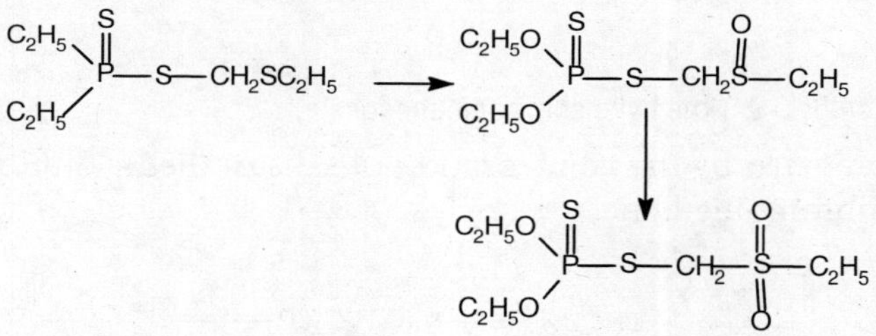

## Disyston

(Disulfoton, O,O-Diethyl-S-(2-ethylmercaptoethyl) dithio phosphate)

Disyston is prepared by treating salts of diethyldithiophosphoric acid with 2-chlorodiethyl sulphide.

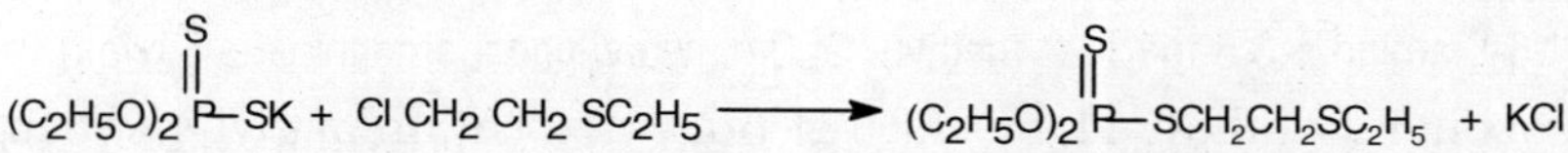

Disyston is a systemic insecticide and is used for control of sucking insects. A disadvantage of disyston is its high toxicity to warmblooded animals. Its $LD_{50}$ for rats is 2-12 mg/kg. Disulfoton is oxidised to sulphoxide and then to sulfon.

The sulfoxide of disyston is marketed as oxydesulfoton. Its biological properties are similar to disyston.

## Phosalone

(O,O-Diethyl-S-(6-chlorohenoxozolinyl-3-methyl) dithio phosphate)

Phosalone is prepared by the condensation of sodium diethyldithiophosphate with 6-chloro-3-chloromethyl benzoxazolone.

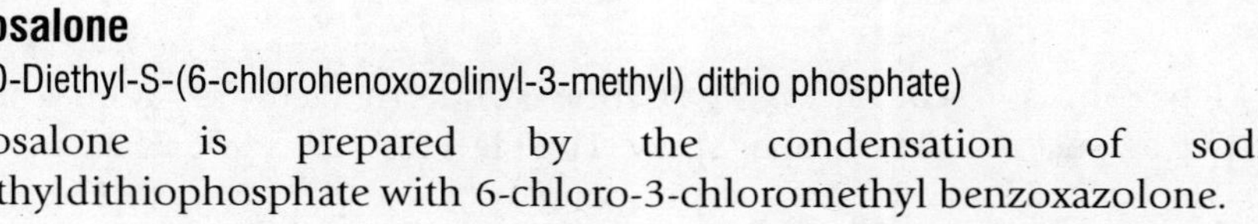

Phosalone is a broad spectrum insecticide and acaricide useful for control of caterpillars, aphids, and mites on apples and field crops. The acute oral $LD_{50}$ value for rats is 120 mg/kg.

## Formothion

(O,O-Dimethyl-S-(N-methyl-N-formylcarbamoylmethyl phosphorodithioate)

Formothion is prepared by the reaction of potassium salt of dimethyl dithiophosphoric acid with N-formyl N-methyl chloroacetamide.

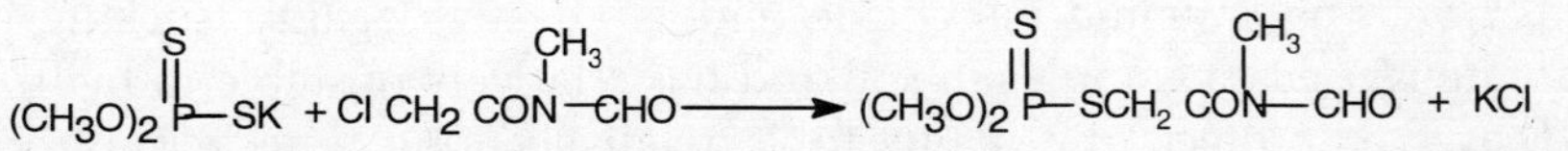

Formothion is yellowish liquid. Formothion has relatively low toxicity to mammals. Its $LD_{50}$ for rats is 330-50 mg/kg and is effective against chewing and sucking pests by systemic and contact action.

## Ethion

(O,O,O,O-tetraethyl -S,S-methylenebis dithiophosphate)

Ethion is prepared by the condensation of salts of diethyldithiophosphoric acid with dibromomethane.

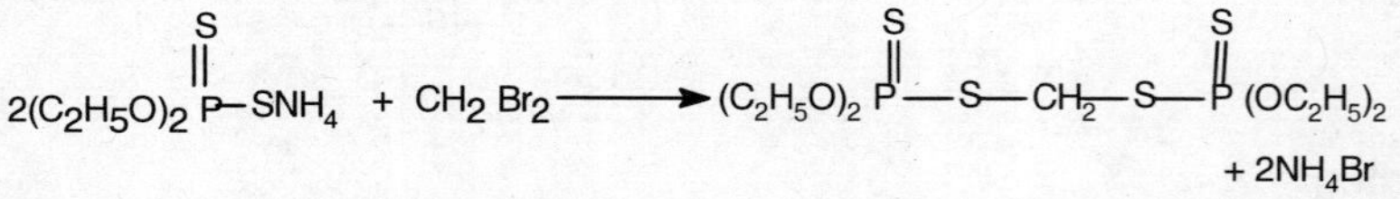

Ethion is a yellow liquid, b.p. 164-165°C. Ethion is used for the control of mites and aphids. The $LD_{50}$ for rats is 55 mg/kg.

## Menazon

(S-(4,6-Diamino-1,3,5-triazin-2-methyl), O, O dimethyl phosporodithioate, Sayfos)

Menazon is synthesised by the reaction of 2-chloromethyl-4,6-diamino-1,3,5-triazine with sodium salt of O,O-dimethyl S-ethoxycarbonyl methyl phosphorodithioate.

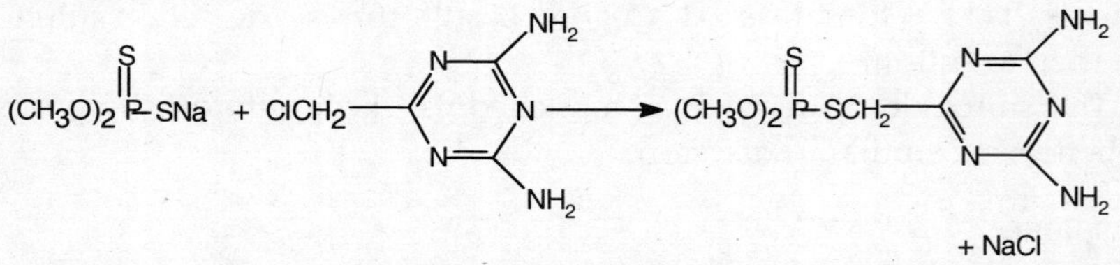

Menazon is a colourless crystalline solid, m.p. 60°C. Menazon is effective against aphids by foliar spray. The $LD_{50}$ for rats is 900-1950 mg/kg.

## Ekatin

(Thiometon, O,O-Dimethyl-S-2(ethylmercaptoethyl) dithio phosphate)

Ekatin is produced by reacting 2-hydroxyldiethyl sulphide with dimethyldithiophosphoric acid and p-toluene sulphonyl chloride in the presence of sodium hydroxide.

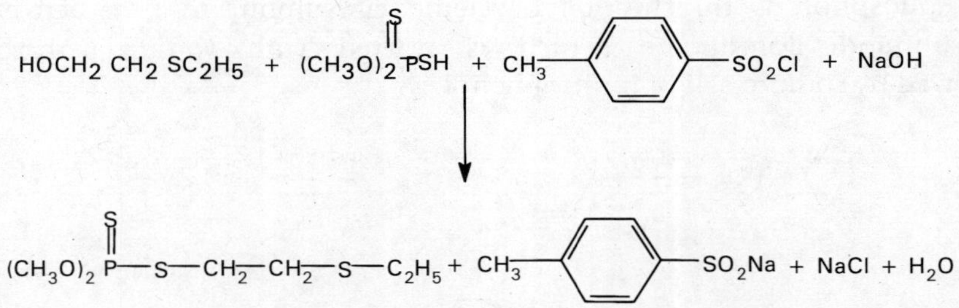

Ekatin is a colourless oil with b.p. 104°C. Ekatin is a systemic insecticide and acaricide with contact activity. It is effective against aphids, thirps and mites. The $LD_{50}$ for rats is 70 to 120 mg/kg.

## Guthion

(Azinphos, 0,0-dimethyl-S-(3,4-dichloro-4-keto-1,2,3-benzotriazinyl-3-methyl) dithiophosphate)

Guthion is prepared by treating N-halo methyl derivative of azimidobenzoyl with salts of dimethyl dithiophosphoric acid.

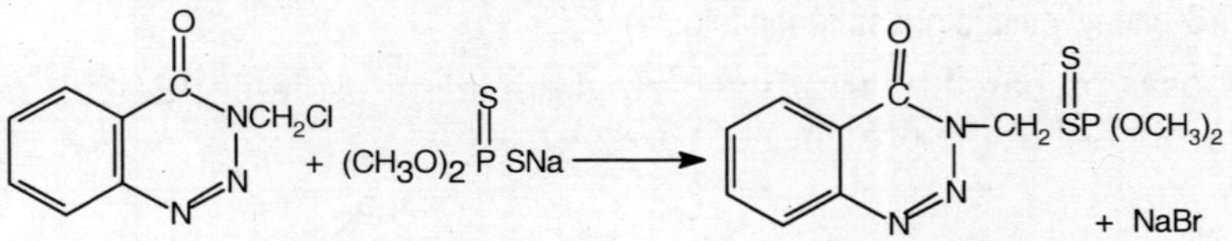

Guthion is white crystalline substance, m.p. 73 to 74°C. It is non sytemic insecticide and acaricide for use in fruit crops, vegetables and cotton. The $LD_{50}$ value for guthion is 15 mg/kg.

## Cidial

(Phenthoate, 0,0-dimethyl-S- (1-carboethoxy benzyl) dithio phosphate)

Cidial is prepared by reacting sodium salt of dimethyl dithiophosphate with ethyl α-bromo phenylacetate.

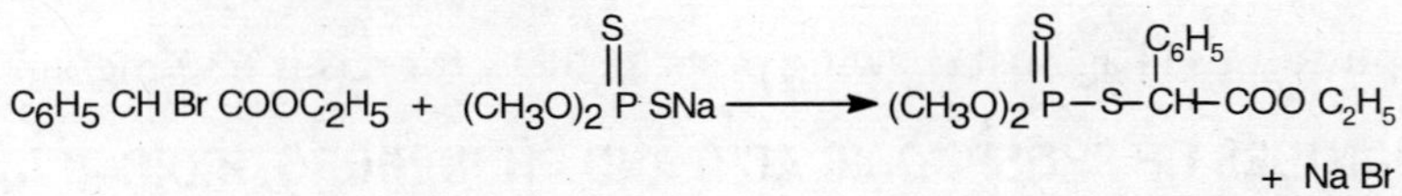

Cidial is a liquid with b.p. 70 to 80°C at 2 x 10-5 mm Hg. Cidial has a broad spectrum of insecticidal and acaricidal activities and is particularly effective against scale nsects. The oral $LD_{50}$ value for rats is 4700 mg/kg.

# PHOSPHONIC ACID DERIVATIVES

## EPN

(0,ethyl 0-(4 nitrophenyl) phenyl phosphonothioate)

Benzene is reacted with phosphorous trichloride in the presence of aluminium chloride to give benzene phosphonodichloridate. This is reacted with phosphorous thiochloride or elemental sulphur to give benzene-phosphonodichlorothioate, which is reacted with sodium ethoxide followed by sodium salt of p-nitrophenol.

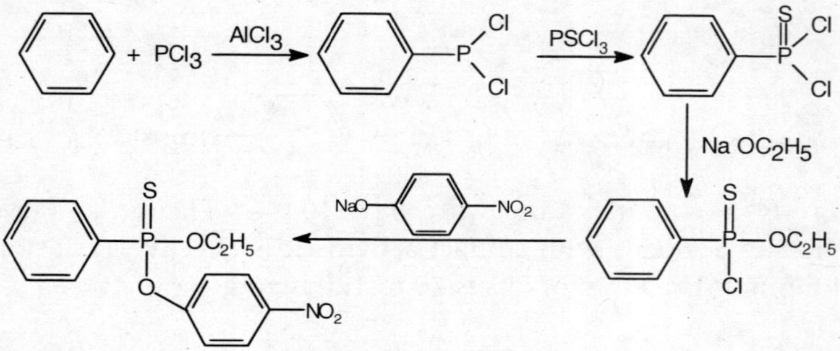

The m.p. of EPN is 36°C. The technical material is a brown liquid. The commercial product has relatively low toxicity against spider mites, grasshoppers and mosquito larvae. The $LD_{50}$ for rats is 36 mg/kg.

### Fonophos

(O-ethyl-S-phenyl ethane phosphonodithioate)

Fenophos is prepared by treating O-ethyl-ethane phosphonothio-chloridate and thiophenol in the presence of base.

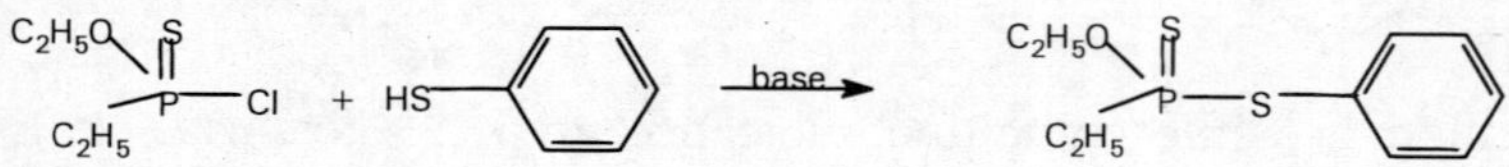

### Diptrex

(O,O-Dimethyl-(1-hydroxy-2,2,2-trichloro) ethyl phosphonate)

Diptrex is prepared by the addition of dimethylphosphonate to chloral

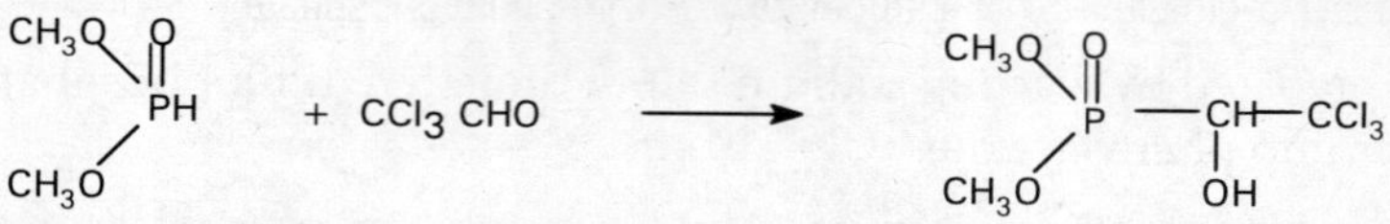

Diptrex is used against chewing pests. Its $LD_{50}$ for rats is 630 mg/kg.

## ESTERAMIDES OF PHOSPHORIC ACID AND PHOSPHOROTHIOIC ACID

### Phospholan

(O,O-Diethyl-N-(1,3-dithiolan-2-ylidene) phosphoromidate)

Phospholan is prepared by the reaction of ethylene 1,2-dithiol with cyanogen chloride, followed by the reaction of 2-imino-1,3- dithiolane with diethyl phosphorochloridate.

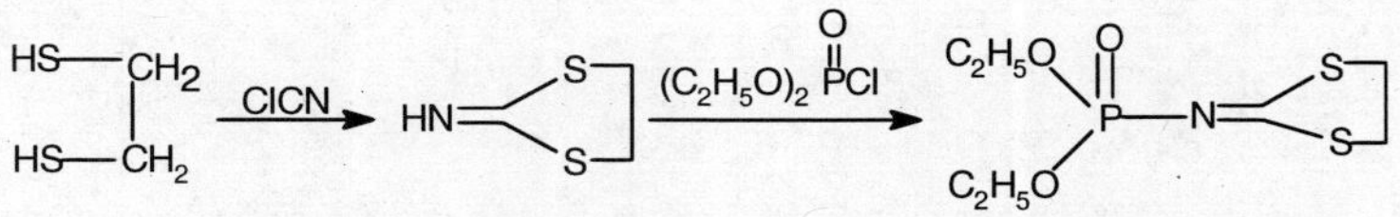

Phospholan has high activity against chewing and sucking insects.

## Methamidophos

(Tamaron, O,S-Dimethyl phosphorothioamidate)

Methamidophos is prepared by converting dimethyl phosphorothiochloridate with ammonia to O,O-dimethyl phosphorothioamidate which in the presence of methyl iodide undergoes thiono-thiolo isomerisation and is converted to O,S-dimethyl analogue.

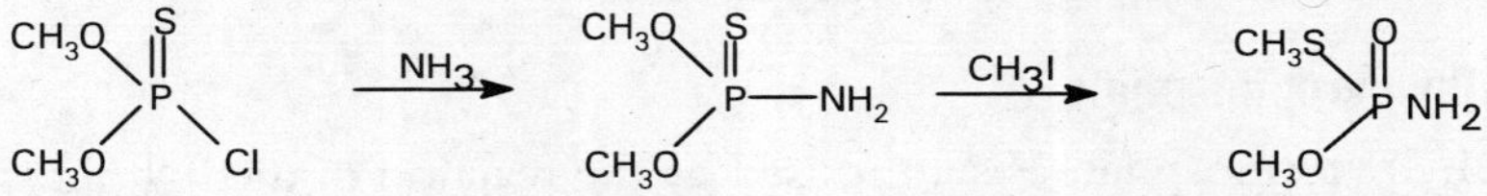

Methamidophos is a colourless crystalline solid, m.p. 44.5 °C. Methamidophos is a broad spectrum insecticide particularly effective to control caterpillers and aphids. It has also systemic action against both biting and sucking pests. Its acute oral $LD_{50}$ value for rats is 30 mg/kg.

## Acephate

(Orthene, O,S-Dimethyl N-acetyl phosphoramidothiolate)

Acephate is N-acetyl derivative of the insecticide methamidophos and is prepared by acetylation of methamidophos with acetic anhydride in the preesence of $ZnCl_2$ or $FeCl_3$.

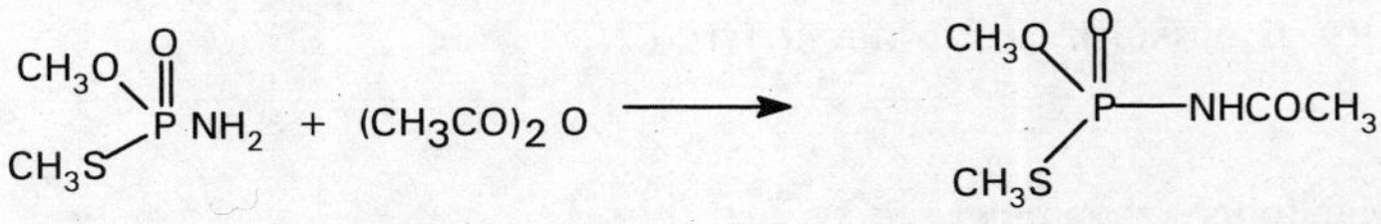

Acephate is a white crystalline solid, m.p. 91 to 92 °C. Acephate is a systemic insecticide effective against chewing and sucking pests. The $LD_{50}$ value for rats is 945 mg/kg.

# CYCLIC PHOSPHATES AND PHOSPHOROTHIOATES

## Salithion

(2-methoxy-4H-1,3,2-benzodioxophosphorin-2-sulphide) and Salioxon (2-methoxy -4H-1,3,2-benzodioxaphosphoran-2-one)

Among the cyclic phosphates, there are also insecticidal derivatives of Salioxon and its thiono analogue salithion. Salithion is prepared by treating saligenin with methyl phosphorodichlorothioate.

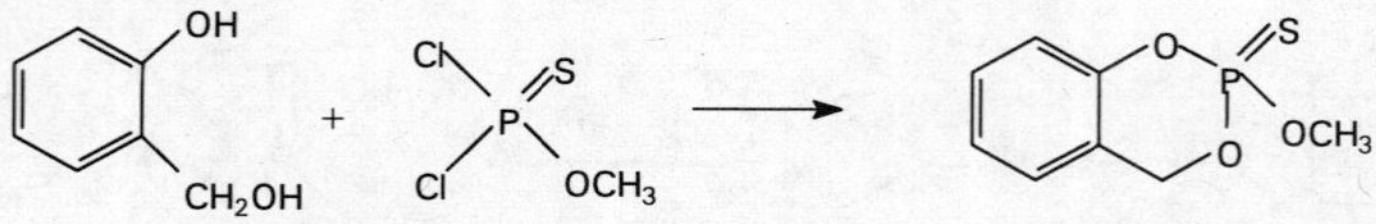

Salithion

The $LD_{50}$ value to rats for salithion is 102 mg/kg and for Salioxon is 0.5 to 1 mg/kg.

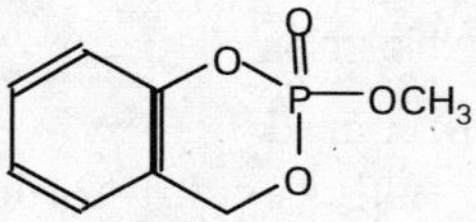

Salioxon

## DERIVATIVES OF PYROPHOSPHORIC ACID

### Tetraethylpyrophosphate

Tetraethylpyrophosphate is synthesised by the reaction of diethyl phosphite with diethyl phosphorochloridate followed by oxidation.

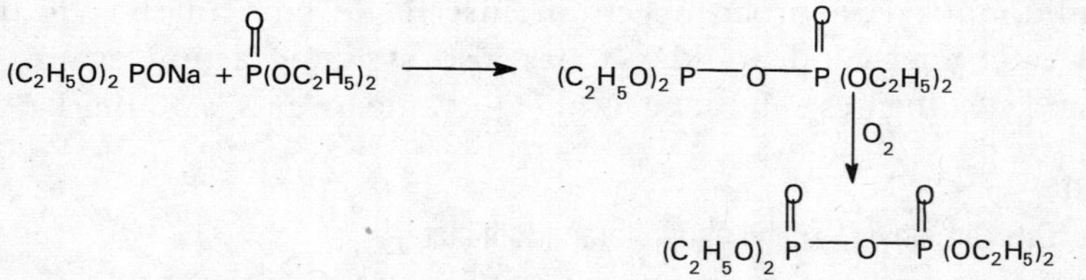

TEPP is colourless, odourless and hygroscopic liquid b.p. 124°C at 1 mm Hg. TEPP is a contact insecticide and acaricide particularly effective to control aphids and mites. The oral $LD_{50}$ for rats is 2.0 mg/kg. Because of high toxicity, the use of TEPP is restricted.

### Schradan

(Octa methyl pyrophosphoamide)

Schradon is prepared by treating dimethylamine with phosphorus oxychloride in the presence of tertiary amine.

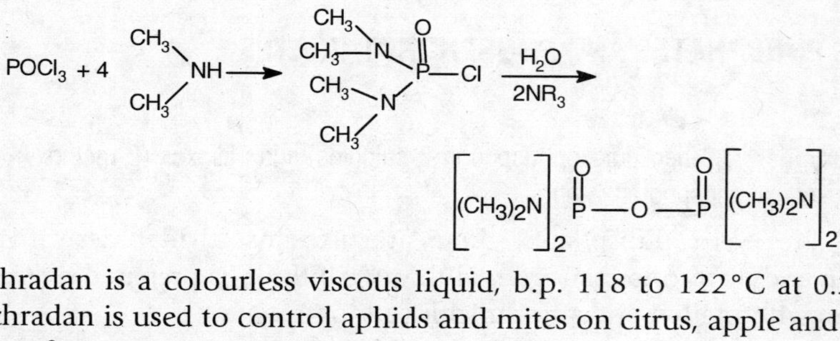

Schradan is a colourless viscous liquid, b.p. 118 to 122°C at 0.3 mm Hg. Schradan is used to control aphids and mites on citrus, apple and hops. The $LD_{50}$ for rats is 35.0 mg/kg.

## MODE OF ACTION OF ORGANOPHOSPHATES AND CARBAMATES

Acetylcholinesterase is an esssential components of nervous systems of both insects and mammals and plays an important role in the transmission of nerve impulses. Acetylcholinesterase catalyses the hydrolysis of acetylcholine to choline and acetic acid.

$$(CH_3)_3\overset{+}{N}CH_2\,CH_2\,OCO\,CH_3 \xrightleftharpoons[H_2O]{\text{acetyl cholinesterase}} (CH_3)_3\overset{+}{N}\,CH_2\,CH_2\,OH + CH_3COOH$$

The organophosphorous and carbamate owe their insecticidal properties because they phosphorylate or carbomylate the enzyme acetyl cholinesterase. In the absence of effective cholinesterase, there is an accumulation of acetylcholine which violates the function of the nervous system. This results in giving rise to the typical chlolinergic symptoms associated in insects with poisoning hyper, activity, tremors, convulsions, paralysis and death. The active centres of the enzyme acetyl cholinesterase contains two main active sites, an anionic site which is negetively charged and binds into the cationic part of the substrate (acetylcholine) and the esteratic site containing the primary alcoholic group of the amino acid serine which attacks the carbonyl carbon atom of the substrate. The normal enzymatic hydrolysis of acetylcholine to choline may be illustrated as follows.

$$CH_3\overset{O}{\overset{\|}{C}}—OCH_2\,CH_2\,\overset{+}{N}(CH_3)_3 + EH \underset{K_{-1}}{\overset{K}{\rightleftharpoons}} CH_3\overset{O}{\overset{\|}{C}}—OCH_2\,CH_2\,\overset{+}{N}(CH_3)_3EH$$

$$CH_3\overset{O}{\overset{\|}{C}}O\,CH_2\,CH_2\,\overset{+}{N}(CH_3)_3\,.\,EH \xrightarrow{K_2} CH_3\,\overset{O}{\overset{\|}{C}}E + HOCH_2\,CH_2\overset{+}{N}(CH_3)_3$$

$$CH_3\,\overset{O}{\overset{\|}{C}}E + H_2O \longrightarrow EH + CH_3COOH$$

The reaction between esterase and phosphorous or carbamate inhibitor is bimolecular SN2 type and can be illustrated as follows.

In comparative studies on acetylcholinesterase inhibitors, only the $I_{50}$ value is measured. In the normal process, step 3 occurs very rapidly, while in the inhibition process 'B' and 'C' the step 3 occurs very slowly because the phosphorous oxygen bond is very strong. So the organophosphate and carbamate effectively poisons the enzyme by phosphorylation or carbomylation, and thus blocks the efficient hydrolysis of acetylcholine into choline. The toxicity of different organophosphates and carbamates is different because the toxicity depends on the phosphorylating or carbomylating constant $K_3$. Thus it has been found that constant $K_3$, the rate constant that determines the regeneration of the enzyme is practically equal to zero for organophosphate insecticides, where as value of $K_3$ for carbamates is small but infinite (when the carbamate group is $CONHCH_3$, K=0.01, when the carbamate group is $CON(CH_3)_2$ K=O.003, when the organophosphate group is $PO(OR)_2$ K=0). The difference between the inhibition of cholinesterase by organophosphate and carbamate is that typical organophosphates are about 30 times more phosphorylating than carbomylating but have lesser affinity between the enzyme and inhibitor. The reversion of inhibited acetylcholinesterase enzyme is more in case of carbamates so the enzyme is activated again and in the case of organophosphate insecticides the reversion from the inhibited enzyme is negligible.

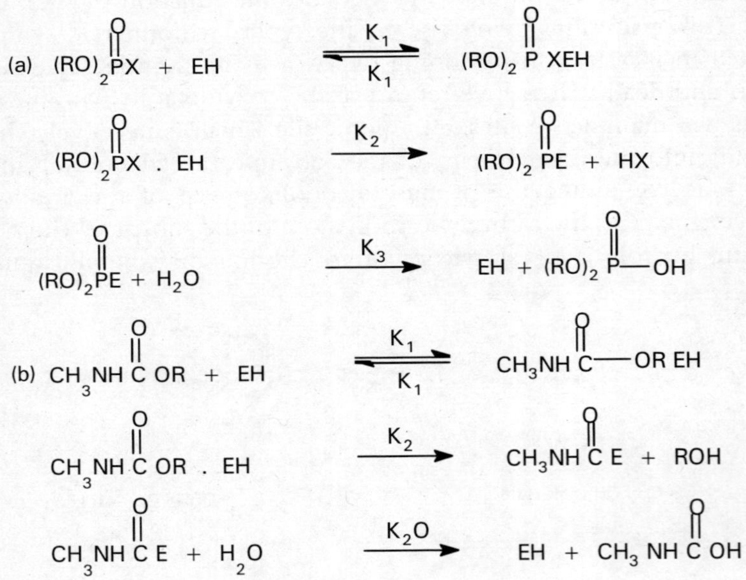

## STRUCTURE ACTIVITY RELATIONSHIP

### Phosphates

The organophosphate type of insecticides are vinyl ester derivatives which generally have an electron withdrawing group such as halogens and a carboxylic ester or amide group on the position. Toxicity depends on the development of +ve charge on phosphorous which is enhanced by the shift of electrons towards the aryl ring. This depends on the dissociation of substituents. The higher the dissociation constant such as alcohol, phenol or acid, the more toxic is the compound. When the length of alkyl radicals in the esters and amides is increased, the toxicity is decreased. The maximum toxicity for many mixed esters of phosphoric acid occurs as diethyl derivatives. Diethyl phosphates are less toxic to insects and mammals. The activity of arylphosphates substituted in position 1 is higher than in position 2 and 3.

## Thiophosphates

Thiophosphates are safer than phosphates from the mammalian toxicity point of view. In thiophosphates, the toxicity depends on the nature of R. $CH_3$, $C_2H_5$ are more toxic than higher alkyl groups. No toxicity of thiophosphates has been observed when the aryl ring is not substituted by the functional groups. The toxicity of thiophosphates is increased if the substituents are -$SCH_3$, $NO_2$, sulphoxide, sulfone, present in para position. When disubstituent is present in 3 and 4 position, toxicity towards mammals is decreased without change in toxicity to insects.

Thiophosphates having the formula

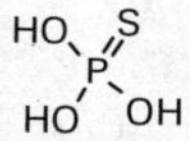


If the carbon atoms in the alkoxy group are more than 4, then activity is decreased. The number of methylene (-$CH_2$-) groups between P and S should not be more than 2, because the activity will be decreased. If more than 3 carbon atoms are present activity is decreased and maximum toxicity is found in 1C. If P is attached to aromatic ring then toxicity is also decreased.

## Dithiophosphates

All derivatives of dithiophosphoric acid are less toxic to mammals than the corresponding derivatives of thiophosphoric acid. All derivatives of dithiophosphoric acid in

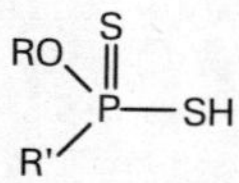


Where R and R' are methyl radicals and are less stable than the corresponding ethyl esters as insecticides. An increase in number of carbon atoms in these radicals increases the toxicity towards mammals without increasing the insecticidal activity. In dialkyl aryl phosphorothionates, maximum activity is obtained, if the substituent group is in the p-position and must be electron attracting. The introduction of a second substituent on the benzene ring such as chlorine or the methyl group reduces mammalian toxicity without any decrease in insecticidal activity. Meta substitution is effective for insecticidal activity. If the aryl group is replaced by aromatic heterocyclics, the insecticidal property is retained but the mammalian toxicity is reduced.

## Pyrophosphoric Acids

The insecticidal activity of pyrophosphoric acid derivative is decreased with increase in size of ester radical. Introduction of second sulphur atom in phosphorous lowers toxicity of the compound to mammals without reducing the insecticidal activity and introduction of a third sulphur on the phosphorous decreased the insecticidal activity.

### Phosphoric Acids

In phosphoric acids, the toxicity of methyl and ethylphosphoric acids to mammals is higher than that of analogues of phosphoric acids. The derivatives of thiophonic acid are less toxic than thiophoric acids. An increase in size of the hydrocarbon radical connected with phosphorous lowers the insecticidal activity.

### Cyclic Esters

In cyclic ester derivatives of phosphates, five membered cyclic phosphate esters are very reactive but these are only poor inhibitors of acetylcholinesterase. On the other hand, five memberd cyclic phosphoroamidates have moderate action on anticholinesterase and insecticidal activity. Six membered cyclic phosphate esters are not so unstable and have poor activity. The poor activity may be attributed due to steric effects of the cyclic structures.

### Asymmetry

Toxicity is also affected by the asymmetry. The leavo isomer of O-ethyl S-2 (ethyl thio)-ethyl phosphonothiolate is ten times more toxic than dextro isomer. The leavo isomer of O,S-dimethyl p-nitrophenyl phosphorothiolate is about five times toxic than dextro isomer. Asymmetry in the succinate moiety of malaoxon also affects inhibitors activity on esterases. d-Malaoxon is about 4 to 8 times better than l-isomer in inhibition of acetylcholinesterase.

## BOTANICAL PESTICIDES

### Nicotine

(3-(1-methyl-2-pyrrolidyl) pyridine)

Nicotine is an alkaloid present in the leaves of Nicotiana tabacum, Nicotiana glutinosa and Nicotiana rustica.

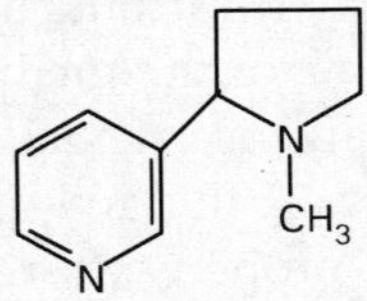


Nicotine occurs in tobacco plants as a salt with citric and malic acid to the extent of 1-8% and may be extracted from leaves and roots of the plants by treatment with aqueous alkali followed by steam distillation. It is converted to nicotine sulphate by treatment with $H_2SO_4$, because it is stable. Nicotine sulphate is being widely used as an insecticide because it is much less toxic to warm blooded animals than free nicotine and is less volatile. Nicotine is commonly formulated as 3.6% dust. Its use is rapidly declining because of its high mammalian toxicity ($LD_{50}$ for rats 50 mg/kg), b.p. 281 °C.

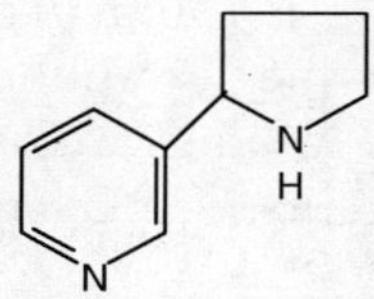

Nornicotine occurs as both d and l forms, the former is present in *Duboisia hopwoodii* and later predominating in *Nicotiana sylvestris*.

Anabasine occurs in the Tuskestan weed *Anabasis aphylla* and has shown aphicidal activity.

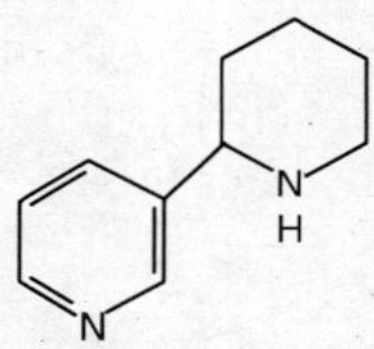

## STRUCTURE ACTIVITY RELATIONSHIP

The essential requirements for toxicity of nicotinoids are a) pyridine ring 2) a highly basic nitrogen c) an unsubstituted α- position of the pyridine ring d) The basicity of the compound is very important, PKa values for the pyridinyl nitrogen moiety (PKa) should be higher than 5.8, preferably 7.4-9.0) The minimum distance between the two nitrogens is about 4.2 A° and this distance is in approximate agreement with that of the two functional groups of the acetylcholine molecule.

### Rotenone

Rotenone occurs in the roots of *Derris elliptica* (from the East Indies and Malya) and a species of *Lonchocarpus utilis* (from South America). Rotenoids can be extracted from the powdered roots with organic solvents.

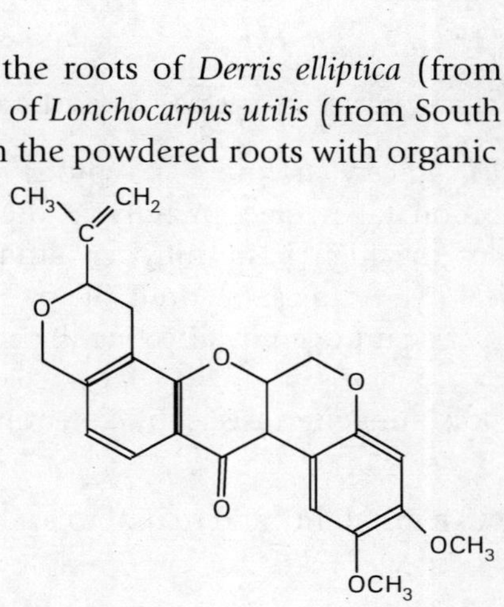

Rotenone is a white to yellowish white crystal, melting point 163 °C, and is soluble in polar solvents and insoluble in water. It is readily detoxified by the action of air and light. Almost all toxicity may lost after 2-3 days of summer exposure.

The $LD_{50}$ of rotenone to rats is 150 mg/kg. It is used as fish poison. It is formulated as dusts. With the exception of hydrogenation of the isopropenyl side chain which produces dihydrorotenone of nearly equal toxicity all other chemical alternations greatly decrease the toxicity of the rotenone molecule.

## Mode of Action

***Nicotine:*** Symptoms of poisoning by nicotine follow the general sequence of (1) excitation, (2) convulsion, (3) paralysis and (4) death. Nicotine stimulates the rate of heart beat at low concentration and decrease it at high concentrations. The onset of symptoms from nicotine poisoning in insects is described as (1) period of inactivity, (2) sequence of paralysis begining with hindlegs, then wings and other legs, (3) toppling of the insect on its back, and (4) total paralysis.

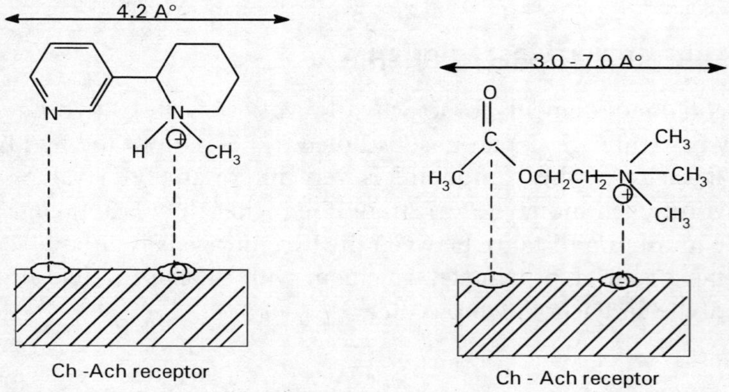


**FIG 4.1:** Similarities between acetylcholine and nicotine binding to the acetylcholine receptor

Nicotine attacks acetylcholine receptors, which resemble cholinesterases. Inhibition takes place mainly at the anionic site by the competitive binding of the high basic nitrogen atom which is normally ionized at physiological pH. At the same time, the pyridinyl nitrogen assists the action of the high-basic nitrogen by affecting the electronic status of the receptor protein.

**Rotenone:** The poisoning of insects by rotenone is usually a slow process:

1. inactivity, locomotive instability and refusal to eat
2. knockdown
3. paralysis

4. slow death.

Rotenone poisoning mainly derives from the ability of rotenone to inhibit respiratory metabolism. More precisely, rotenone owes its inhibitory potency to its ability to interfere with the electron-transport process between reduced diphosphopyridine nucleotide (DPNH or NADH) and cytochrome b. Inhibition of respiratory metabolism is one of the major causes of nerve conduction block.

## Pyrethrins

Pyrethrum is derived from dried flowers of the plant *Chrysanthemum cinerariaefolium*. The name given to the active insecticidal components of the dried flowers is known as pyrethrins. Chemically "pyrethrins" are organic esters formed by the combination of two carboxylic acids and three keto alcohols. One set of esters is known as the pyrethrin group consisting of pyrethrin I, a combination of chrysanthemic acid with pyrethrolone and pyrethrin II formed by the combination of pyrethric acid with pyrethrolone.

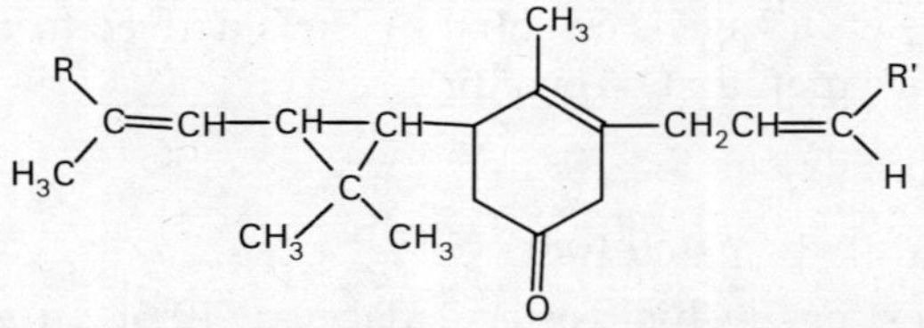


| | R | R |
|---|---|---|
| Pyrethrin I | $-CH=CH_2$ | $-CH_3$ |
| Parethrin II | $-CH=CH_2$ | $-COOCH_3$ |
| Cinerin I | $-CH_3$ | $-CH_3$ |
| Cinerin II | $-CH_3$ | $-COOCH_3$ |
| Jasmolin I | $-CH_2\ CH_3$ | $-CH_3$ |
| Jasmolin II | $-CH_2CH_3$ | $-COOCH_3$ |

The second group is known as cinerin esters. These consist of cinerin I and cinerin II and are formed by combination of cineralone with chrysanthemic acid and pyrethric acid respectively. The third group was discovered in 1964 and is known as Jasmoline esters. These consist of Jasmoline I and Jasmoline II and are formed by the combination of jasmalolone with chrysanthemic acid and pyrethric acid respectively.

The active consistuent pyrethrins from pyrethreum flowers are extracted with solvents like kerosene. The solvent is percolated through ground flowers and extract evaporated under vaccum.

The outstanding properties of pyrethrins are

1. **Rapid action:** It is a contact insecticide attacking the nervous system of insects almost immediately and causing knockdown followed by kill.

2. **Low mammalian toxicity:** It has a low mammalian toxicity. It is one of least toxic domestic insecticide available.
3. **Broad spectrum of Insecticides:** Because it consists of a group of related compounds, it has a wider spectrum of activity against insect species than any single insecticide.
4. **Lack of persistence:** Pyrethrum is degraded by the combination of sunlight and air, and therefore presents little of the hazards which are usually associated with certain other classes of persistent insecticides.
5. **Repellency:** Pyrethrum is a powerful insect repellent which in combination with its low mammalian toxicity favours it in many applications.

## STRUCTURE ACTIVITY RELATIONSHIP

The most active isomers are d-trans acid esters of the d-cis alcohols. For biological activity presence of a) isobutylene group b) the dimethyl cyclopropane ring of the acid c) the unsaturated keto group of the keto alcohol d) the ester linkage is essential e) esters derived from chrysanthemic acid is more active than that of pyrethric acid.

### Synthetic Pyrethroids

Advantages of synthetic pyrethroids:

5. They have extremely high insecticidal activity at extremely low doses and are biodegradable in nature.
6. Their activity is most pronounced against lepidopterous pests and are also effective against beetles, leaf miners and bugs.
7. They maintain insecticidal activity over a great extended period of time which helps in controlling overlapping generations of pests.
8. They are effective against eggs, larval and adult stages of insects, particularly so in the case of most lepidoptera insects.
9. They have powerful antifeedant and repellent properties, after contact insects become moribund and stop feeding.
10. They have rainfastness i.e. they are not readily washed off from the plants by rain. The lipophilic character enables ready absorption on waxy layers of plant surface.
11. Synthetic pyrethroid treated crops give increased yields of better quality and in the case of cotton, the crop tends to mature early.
12. The movement from site of application is limited in air due to low volatility and in soil and plants due to low polarity.

*Disadvantages*

1. Synthetic pyrethroids are not systemic and do not have translamiar action, so they are not effective against sucking pests like aphids and mealy bugs resulting in the resurgence of these pests.
2. Constant use of synthetic pyrethroids increase incidence of mites in

crops like cotton, tea, vegetables etc.

3. With large scale use of synthetic pyrethroids incidence of Alternaria disease increased in cotton mostly in sandy soil.
4. The synthetic pyrethroids are highly toxic to fish. The lipophilicity of pyrethroids indicates, that they will be absorbed strongly by the gills of fish even at very low concentration in water.

Some of the synthetic pyrethroids do not have a specific antidote.

## Allethrin

(2-methyl-4-oxo-3-(2-propenyl)-2 cyclopenten 1-yl)

Allethrin is prepared by esterification of synthetic ± chrysanthemic acid with the alcohol allethrolone.

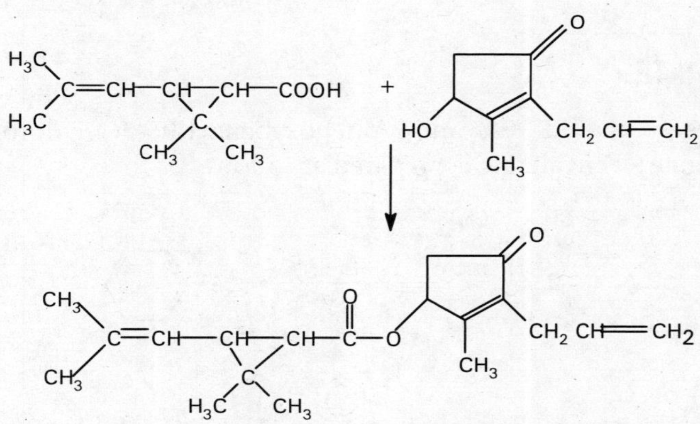


Allethrin is a mixture of four racemic forms or eight optical isomers, four of which are esters of the cis-acid and four are esters of trans acid.

Technical allethrin is a pale yellow liquid b.p. 140°C /0.1 mm Hg. Allethrin is a contact insecticide effective against house hold pests. The acute oral $LD_{50}$ for male rats is 1100 mg/kg and female 685 mg/kg. Bioallethrin is D-trans allethrin. It is amber viscous liquid b.p. 153°C/ 0.4 mm Hg. It is a contact non-systemic insecticide which produces a rapid knockdown and is used against house hold pests. The prefix bio is generally used to indicate the esters of the natural dextrorotatary (1R)-trans-chrysanthemic acid.

## RESMETHRIN AND BIORESMETHRIN

Resmethrin (5 benzyl-3 furanyl methyl (1RS)-cis-trans-2,2-dimethyl-3-(2-methylprop-1-enyl) cyclopropane carboxylate) is prepared by treating chrysanthemic acid with 2-benzyl-4-furylmethyl alcohol.

Resmethrin is a colourless waxy solid m.p. 43-48°C. It is used to control agricultural, horticultural and public health insect pests. The $LD_{50}$ for rats is more than 2500 mg/kg. Bioresmethrin is the ester of the (1R)-trans acid. It is a yellow liquid m.p. 30-35°C, b.p. 181°C/0.1 mm Hg. It is also effective against wide range of agricultural and house hold pests. The acute oral $LD_{50}$ for rats is 7070-8000 mg/kg.

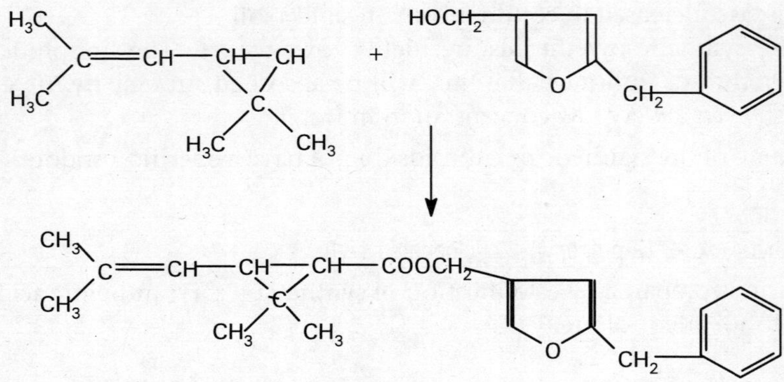


## FENVALERATE

Fenvalerate ((RS)-α-cyano-3-phenoxybenzyl-RS-2(4-chlorophenyl)-3-methylbutrate). Fenvalerate is prepared as follows

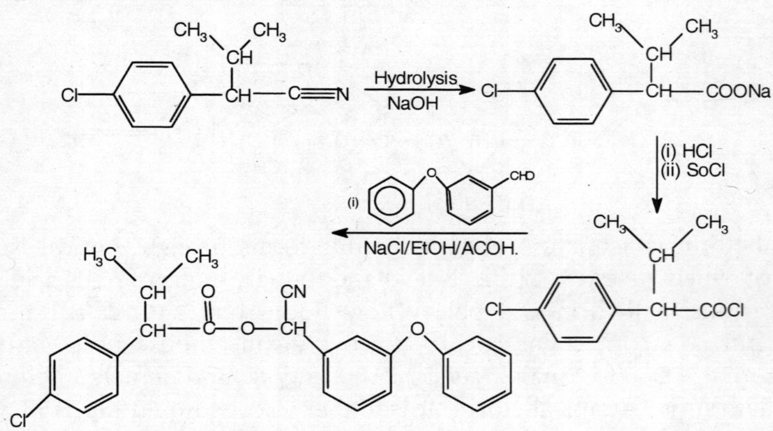


Fenvalerate is a viscous yellow liquid. Fenvalerate acts both as a contact and stomach poison. It controls the pests on crops of cotton, vegetables and fruits. Its $LD_{50}$ for rats is 451 mg/kg.

## PERMETHRIN

Permethrin (3-phenoxy benzyl (1RS) cis, trans-3 (2,2- dichlorovinyl) 2,2-dimethylcyclopropane carboxylate) is prepared by converting methyl (1RS)-cis, trans-3 (2,2 dichlorovinyl)-2,2-dimethyl cyclopropane carboxylate to corresponding acid chloride and then reacting this intermediate with 3-phenoxy benzyl alcohol.

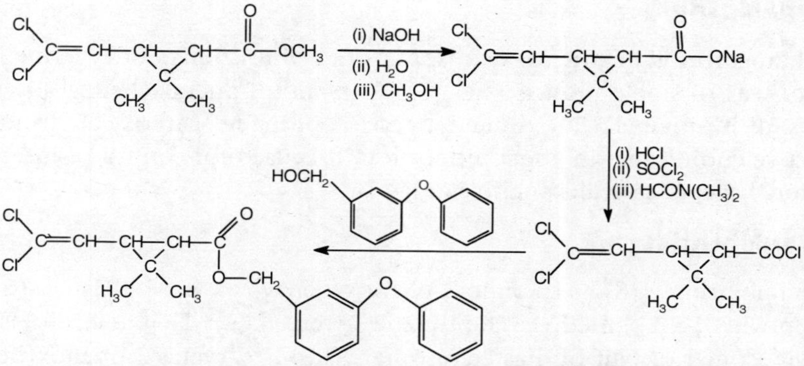


Permethrin is a white, odourless, crystalline solid. The melting point of cis isomer is 54-56°C and trans isomer is 45-46°C. The boiling point of permethrin is 210-220°C at 0.05 mm Hg. Permethrin is a contact insecticide effective against broad range of pests of cotton, fruits, vegetables and tobacco. The acute oral $LD_{50}$ values for *cis*: trans ratio of about 40:60 are 430 to 4000 mg/kg.

## CYPERMETHRIN

Cypermethrin ((RS) - α-cyano-3-phenoxy benzyl (1RS) - cis, trans-3(2,2-dichlorovinyl)-2,2-dimethylcyclopropane carboxylate) is prepared by converting methyl (1RS)-cis, trans-3-(2,2-dichlorovinyl) 2,2-dimethylcyclopropanecarboxylate to the corresponding acid chloride and then reacting this intermediate with 3- phenoxybenzaldehyde in the presence of sodium cyanide.

Pure cypermethrin is a white, odourless, waxy solid. The m.p. of pure cypermethrin isomers is 81-83°C. Cypermethrin is a stomach and contact insecticide effective against broad range of pests of cotton, fruit and vegetable crops. The chemical is active against eggs and active growth stages particularly larvae and also has antifeeding action. The $LD_{50}$ values of cypermethrin is 303 - 4123 mg/kg for rats.

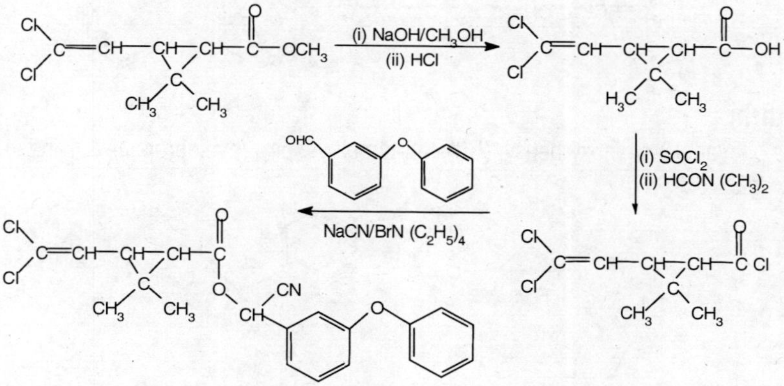

## ALPHAMETHRIN

Alphamethrin is mixture of two isomers out of 8 isomers of cypermethrin. It consists of single most isomer (S) - α-cyano- 3-phenoxybenzyl (1R)-cis-3(2,2-dichlorovinyl)-2,2 dimethyl cyclo propane carboxylate and its inactive enontiomer in equal proportion. It is used for control of insects of cotton, vegetables, fruits and in storage.

## DELTAMETHRIN

Deltamethrin ((S)- α-cyano-3-phenoxybenzyl (1R, 3R) -3-(2,2-di bromovinyl)-2,2-dimethylcyclopropane carboxylate). Deltamethrin is the single isomer of pair of diastereoisomers (RS)- - α-cyano-3-phenoxybenzyl (1R, 3R)-3-(2,2-dibromovinyl) 2,2-dimethyl cyclopropane carboxylate having 5 configurations at the benzyl ring.

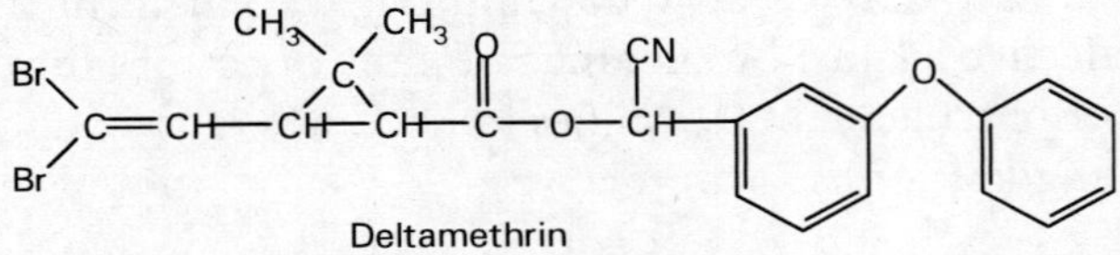

Deltamethrin

Deltamethrin is a colourless crystalline powder m.p. 98-101°C. It is a potent insecticide effective as a contact and stomach poison, against wide range of pests of fruits, vegetables, cotton and of stored products. The acute oral $LD_{50}$ for rats ranges from 135-5000 mg/kg. It is used at low dosages of 12.0 g/ha for the control of pests.

## CYHALOTHRIN

Cyhalothrin ((RS)-α-cyano-3-phenoxybenzyl (±) cis, trans-(2VE)-2-chloro-3,3,3-trifluoroprop-2-enyl)-2,2-dimethylcyclopropane-2- carboxylate)

Cyhalothrin is a colourless solid m.p. 49.2°C. It is effective against pests of cotton, vegetables and fruits. Its $LD_{50}$ for male rats is 244 mg/kg.

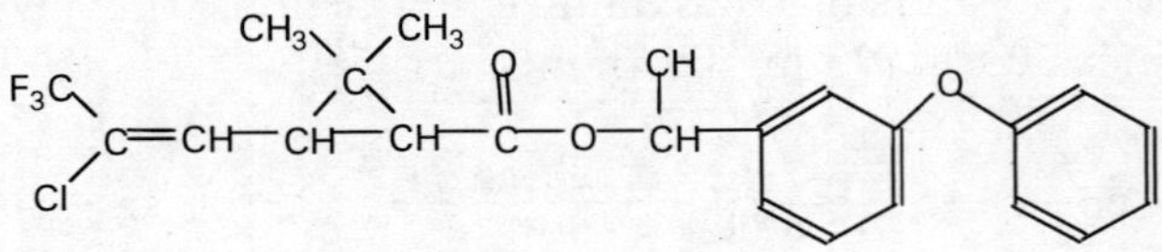

### Cypothrin

((RS) - α-cyano-3 phenoxy benzyl (RS)-3,3-dimethyl spiro (cyclopropane 1,1-indene)-2 carboxylate)

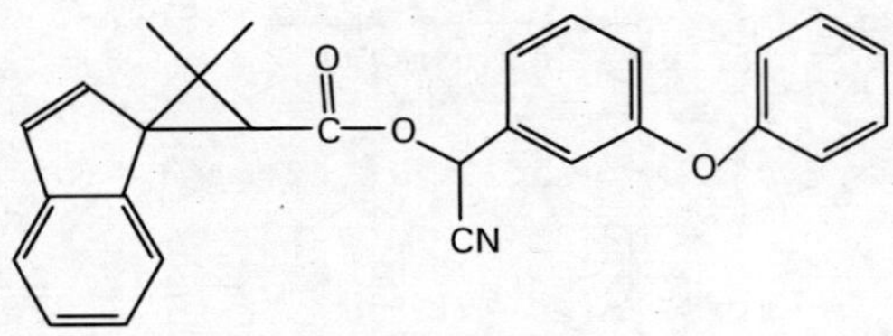

Cypothrin is mainly used in animal health as tickicide.

## Flucythrinate

(α-cyano-3-phenoxybenzyl (S) -2-(4-fluoromethoxy phenyl)-3-methyl butyrate)

Flucythrinate is effective in suppressing phytophagous mites and is used for control of insects on cotton and vegetables.

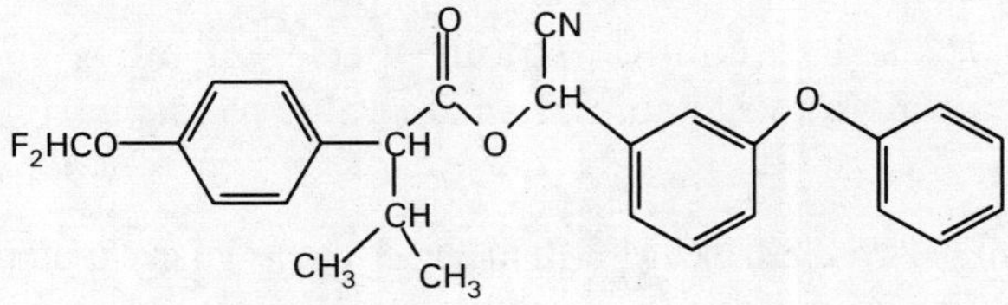

## Fluvalinate

(RS- α-cyano-3-phenoxy benzyl (R-2-(2- chlorotrifluoromethylanilino)-3-methylbutyrate)

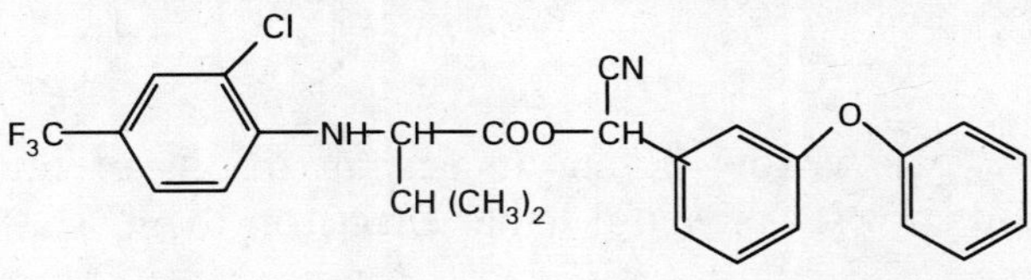

Fluvalinate is viscous yellow oil b.p. 450°C. Fluvalinate is used as a insecticide on cotton, vegetables and ornamental crops. It is also used as an acaricide and aphicide. Its $LD_{50}$ to rats is 250 mg/kg.

## Cyfluthrin

(RS)-α cyano-4-fluorovinyl 2,2-dimethyl cyclopropane carboxylate)

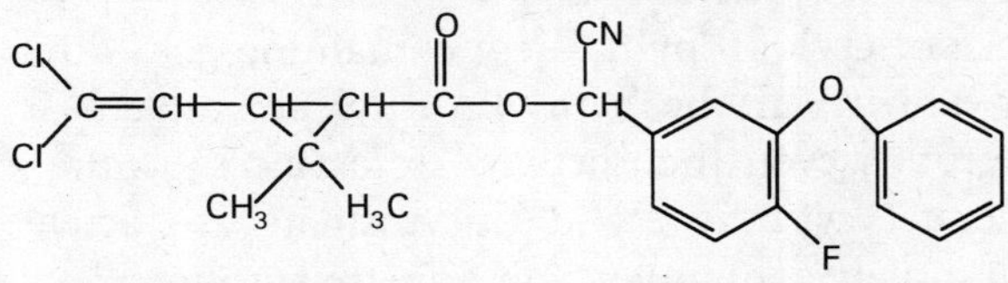

Cyfluthrin is a yellow mass m.p. 60°C. Cyfluthrin is effective against house hold pests. Its $LD_{50}$ value for rats is 500-800 mg/kg.

## Tralomethrin

((S)- α-cyano-3-phenylbenzyl (1R, 3S) -2-2 dimethyl -3-(RS)-1,2,2,2,-tetrabromoethyl cyclopropane carboxylate)

Tralomethrin is effective in controlling pests of cotton, soybeans, potato and vegetables. Its $LD_{50}$ value to rats is 3000 mg/kg.

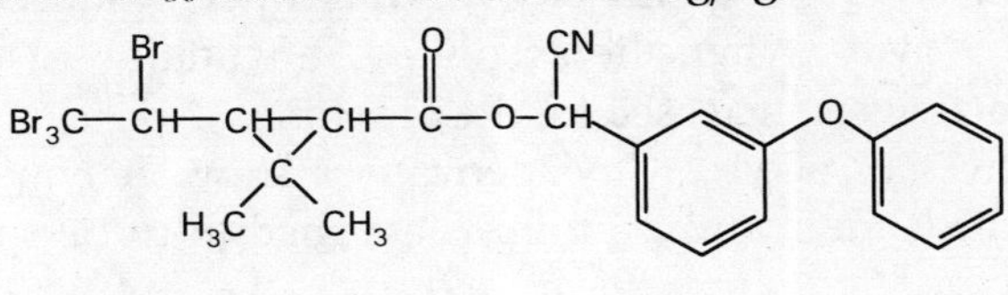

## Fenpropathrin

((R,S)- α-cyano-3-phenoxybenzyl 2,2,3,3-tetramethylcyclopropane carboxylate)

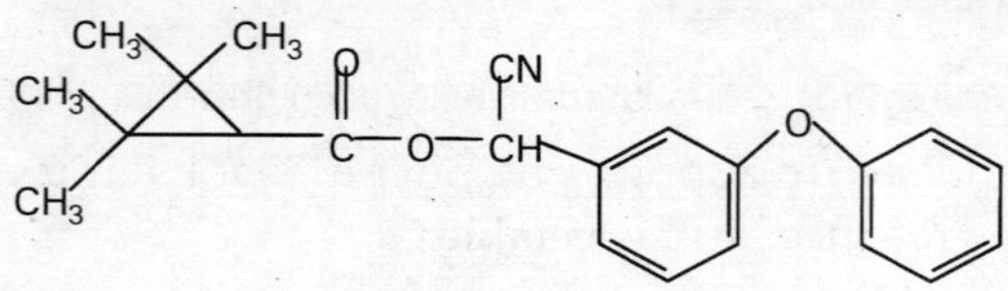

Fenpropathrin is used to control various species of mites and insects on ornamentals, cotton and vegetables. Its $LD_{50}$ value to rats is 164 mg/kg.

## Sumethrin

(3-phenoxyphenyl) methyl 2,2-dimethyl-3-(2 methyl-1-propenyl) cyclo-propane carboxylate

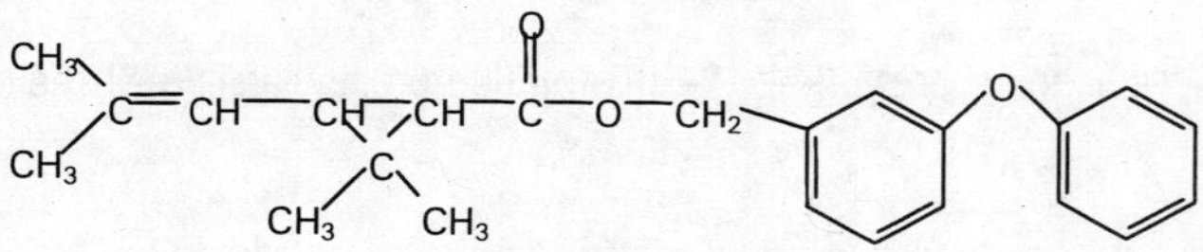

Sumithrin is a pale yellow liquid. It acts on insects of human health importance and shows a powerful lethal effects on them. The $LD_{50}$ values for rats is 5 g/kg.

## STRUCTURE ACTIVITY RELATIONSHIP

1. Pyrethroids are essentially ester compounds. Attempts to replace the ester with other linkages had reduced the insecticidal activity. There is now a single group which falls outside this category. These compounds are ethers of alkyl aryl ketooximes.
2. The presence of cyclopropyl moiety containing gem dimethyl group is crucial for activity. If the gem dimethyl substituent is replaced by spiroalkyl groups even then activity is retained. Synthetic pyrethroids like fenvalerate, cyfluthrin and flucythrinate are active without the cyclopropyl moiety containing gem dimethyl group.
3. Pyrethroids having trans configuration at C-1 and C-3 position of the cyclopropane ring are more active than their corresponding cis isomers except deltamethrin.
4. Two centres of unsaturation at the extremes of the molecule are required for insecticidal activity.
5. The configuration at C-2 and C-3 must either 1R, 3S or 1R, 3R.
6. Esters of acids with 3-phenoxybenzyl alcohol have insecticidal activity. Introduction of cyano group (-CHCN)) further produced changes in insecticidal activity. When the absolute configuration of the -CH(CN) group is S, activity is increased to fifteen fold.
7. The derivatives of alkyl or aryl derivatives show activity. The carbon atoms to which the hydroxy group is attached must be $sp^3$ hybridised.

Phenyl esters do not show activity.

8. Replacement of isobutenyl to dihalovinyl in the acidic moiety increases toxicity especially with difluoro compounds.

## MODE OF ACTION OF SYNTHETIC PYRETHROIDS

There is a general agreement that synthetic pyrethroids owe their insecticidal activity because of their attack on the insect nervous system. The symptoms of pyrethroid poisoning follow the typical pattern of nerve poisoning, namely excitation, convulsions, paralysis and death. On the basis of the mode of action, synthetic pyrethroids can be grouped into two categories, Type I compounds include pyrethrins allethrins, resmethrin, phenothrin, permethrin and oxime ethers. The type II pyrethrins include cypermethrin fenvalerate, deltamethrin. The type I pyrethroids cause restlessness in co-ordination, hyper activity followed by paralysis. Type II causes convulsions hyper activity and contractions. It has been established that pyrethroids are good inhibitors of Ca-Mg and Co-ATPase enzyme systems which in turn disturb Na-Ca exchange and calcium pumping which are necessary to maintain gradient across cellular membranes in the nervous sytem.

## REFERENCES


1. Brown, A.W.A., Insect Control by Chemicals, 817 pp, Wiley, New York, 1951.
2. Colhoun, E.H., J. Agr. Food Chem. 8: 252 (1960)
3. Corrigan, J.J., and Kearns, C.W., J. Insect Physiol. 9: 1 (1968)
4. Dale, W.E., Gaines, T.B., Hayes, W.J., and Pearce, G.W., Science 145: 1052. (1963)
5. Gianotti, O, Metcalf, R.L., and March, R.B., Ann. Entomol. Soc. Am. 49: 588 (1956)
6. Guthrie, F.E., J. Econ. Entomol., 43: 559 (1950)
7. Hart, L.G., Shultice, R.W., and Fouts, J.R., Toxicol. Appl. Pharmacol. 5; 371-86 (1963)
8. Harvey, G.T. and Brown, A.W.A. Can. J. Zool., 29: 42(1951)
9. Kearns, C.W., Quoted in Spencer, E.Y., and O'Brien, R.D., Ann. Rev. Entomol. 2: 261 (1957)
10. Lalonde, D.I.V., and Borwn, A.W.A., Can. J. Zool. 32: 74 (1954).
11. Lewis, S.E., Nature 172: 1004 (1953)
12. Matsumura, F., and O'Brien, R.D., J. Agr. Food Chem. 14: 36(1966)
13. Mullins, L.J. Science 122: 118(1955)
14. Orser, W.B., and Borwn, A.W.A., Can. J. Zool. 29: 54 (1951).
15. Roeder, K.D., and Weiant, E.A., Science, 103: 304 (1966)
16. Shafer, G.D., Mich. Agr. Exp. Sta. Tech. Bull. 11: 65 (1911)
17. Sternburg, J., and Kearns, C.W., Ann. Entomol. Soc. Am. 52: 386 (1952).
18. Tobias, J.M. Kollross, J.J., and Savit, J.J. Cellular Comp. Physiol, 28: 159 (1946)
19. Welsh, J.H., and Gardon, H.T., J. Cellular Comp. Physiol. 30: 147 (1947)
20. Yamasaki, T., and Narahashi, T., Botyur-Kagaku 22: 296 (1957).
21. Yeager, J., and Munson, S., Science 102: 305 (1945).


## BIBLIOGRAPHY


1. Baker D.R., Fenyes J.G., Motiergano W.K., Cross B. Synthesies and chemistry of agrochemicals, American Chemical Society 1987.
2. Shepsrard S.H.K. The Chemistry and action of insecticides, Mae-Grow- Hill, New

York, 1951.
3. Metcalf R.L., Organic Insecticides: Their Chemistry and mode of action Wiley, New York 1955.
4. O'Brien R.D., Insecticides, Action and metabolism, Academic Press, New York, 1967.
5. O'Brien R.D., Toxic Phosphoric Ester Academic Press, New York, 1960.
6. Hassal K.A. World Crop Protection-Pesticides London Ilife Bombs, 1969.
7. Corbett J.R., Wright K. and Baillie A.C. The Biochemical mode of action of Pesticides. Academic Press New York, 1984.
8. Buchel K.H., ed Chemistry of Pesticides, Wiely, New York, 1983.
9. Gy. Matolesy, Nadasy M. and Andorsk V. Pesticide Chemistry, Elserier, 1988.
10. Cropping, L.E, ed crop protection from nature, natural products and analogues, Royal Society of Chemistry, 1966.

# CHAPTER 5

# Fungicides

Chemicals used to kill fungi or control fungi which cause plant diseases are called fungicides. The plant diseases are caused by pathogenic agents known as fungi, viruses and bacteria and parasitic seed plants.

## Fungi

Fungi are plants that lack green colour (chlorophyll). Fungi cannot make their own food. Most fungi reproduce by spores which work about the same way as seeds. Many fungi are capable of penetrating plant tissue directly. Examples of diseases caused by fungi are smut on corn, downy mildew on soybeans and red rot of sugarcane etc.

## Bacteria

Bacteria are microscopic and single celled organisms. They reproduce by simply dividing in half. Each half becomes a fully developed bacterium. Bacteria must enter through natural openings or through wounded tissue. Blight of pears and scab of potatoes are caused by bacteria.

## Viruses

Virus are ultramicroscopic and are seen by electron microscope. They are studied by their effects on selected indicator plants. Virus diseases are transmitted by from one plant to another by insects, cuttings and by mechanically rubbing leaves of healthy plants with juice from diseased plants. Examples of virus diseases are sugarcane mosaic virus and tomato mosaic virus.

## Parasitic Seed Plants

The parasite usually attaches itself to the host and deprives the host of nutrients and water. Mosses and dodders are examples of parasitic seed plants.

Fungicides can be classified on the basis of mode of action, use and chemical constitution. Based on the mode of action these are classified as follows:

A. Fungistatics
B. Protectants
C. Eradicant
D. Systemic Fungicides

1. **Fungistatics:** Chemicals which do not kill the fungi but inhibit the growth of the fungi.
2. **Protectants:** These check entry of the fungi into the plant but do not kill the established organisms.
3. **Eradicants:** These destroy, control or kill the established organisms.
4. **Systemic Fungicides:** A systemic fungicide is a compound that is taken up by the plant and is then translocated within the plant thus protecting it by the attack of pathogenic fungi. For a chemical to be effective systemic fungicide:
   a) It must itself be fungicide or be converted into active fungitoxicant within the host plant
   b) It must have very low toxicity.
   c) It must be capable of absorbed by roots, seeds or leaves of the plant and then translocated at least within the plants.

## CHEMICAL CLASSIFICATION

On the basis of their chemical structure fungicides can be grouped into two major groups 1) Inorganics and 2) Organics

### Inorganics

1. Sulphur fungicides
2. Mercury fungicides
3. Copper fungicides

### Organics

These are organic compounds and are subdivided into various groups as below:

| Organotins | Organomercurials |
|---|---|
| Organophosphorous | Benzene derivatives |
| Diethylthiocarbamates | Polyhalogen alkenic sulphonyl gruop |
| Oxathiins | Quinones |
| Carboxyamides | Dinitro alkylphenols |
| Azole compounds | Benzimidazole derivatives |
| Triazoles | Hetrocyclic fungicides |
| Piperazine derivatives | Pyrimidines |
| Quinoline derivatives | Morpholine derivatives |
| Antibiotics | Miscelleneous |

Classification based on general use is as follow:

1. **Seed treatments:** These are also called seed protectants. These are chemicals which are used for seed or seedling dips for seed borne diseases.
2. **Seed dressings:** These are chemicals which are used for dressing cuts

and treatments of affected stems

3. **Foliage and flower protectants:** These chemicals are used to avoid damage to leaves and flowers.

## REQUIREMENTS OF AN IDEAL FUNGICIDE

An ideal fungicide should have the following characteristics:

1. **Broad spectrum:** An ideal fungicide should have a broad spectrum of activity and it should kill variety of pathogens during the single spray.
2. **Low mammalian toxicity:** A good fungicide should be safe to mammals and beneficial to predators.
3. **Solubility:** A good fungicide should be sufficiently water soluble to give a cheap hydrophilic spray.
4. **Stability:** A good fungicide should be stable during storage and application.
5. **Low fungitoxicity:** It should not be toxic to plants.

## SULPHUR COMPOUNDS

Sulphur is one of the most important fungicides used for the control of various diseases, particularly powdery mildews. Elemental sulphur occurs both in free state and also in combination. Sulphur is used in a finely ground condition or in colloidal form in dust formulations. Ground sulphur is obtainted by grinding coarse sulphur in a micronizers and then fractionating it in with air blasts. As the mechanically ground sulphur tends to aggregate, small amount of inert material like kaolin is added. Most of the sulphur sold for fungicidal use is ground to pass at least a 300 mesh sieve. Sulphur even though finely ground is not wetted by water. Wettable powders of sulphur are made by grinding sulphur with protective colloidal materials such as casein, bentonite clay and flour. Sulphur dusting powders containing materials such as gypsum, bentonite or talc in small amount make the sulphur flow freely and this form is called "conditioned sulphur".

Two forms of sulphur hydrophillic and hydrophobic are also used. Hydrophillic sol is prepared by passing $H_2S$ through a solution of sulphur dioxide in water or by reaction of sodium thiosulphate and sulphuric acid.

$$2H_2S + SO_2 \longrightarrow 3S + 2H_2O$$

$$Na_2S_2O_3 + H_2SO_4 \longrightarrow Na_2SO_4 + SO_2 + S + H_2O$$

Hydrophobic colloidal sulphur is obtainted by acidifying lime sulphur solution with mineral acids.

$$CaSx + 2HCl \longrightarrow CaCl_2 + H_2S + Sx_{-1}$$

Lime sulphur or calcium polysulphide ($CaS_x$) is also used as fungicide. Limesulphur is prepared by boiling 1.8 parts of sulphur, 1 part of lime and 10 parts of water for one hour. The solution is filtered and filtered solution is diluted to specific gravity 1.003, for sprays. This is a clear orange coloured liquid consisting mainly of calcium polysulphides which break down on exposure to air to release elemental sulphur. The polysulphide content of a lime sulphur mixture is the best indication of its effectiveness and the official specification contains at least 24% weight/volume of the polysulphide.

Barium polysulphide ($BaS_x$) is also used as fungicide. It is prepared by saturating barium hydroxide solution with hydrogen sulphide or mixing ground sulphur with barium sulphite power. Calcium tetrathionate is also used as fungicide. It is prepared by passing sulphur dioxide into sulphuryl chloride.

$$2SO_2 + 2H_2O + S_2Cl_2 \longrightarrow H_2SO_4 + 2HCl$$

$$H_2SO_4 + CaCl_2 \longrightarrow CaS_4O_6 + 2HCl$$

It is found to be as effective as colloidal sulphur.

### Mode of Action of Sulphur

The sulphur compounds act as hydrogen acceptor and thus interferes with the normal redox process of the fungal cells.

## MERCURY FUNGICIDES

Amongst the inorganic mercuric compounds mercuric chloride ($HgCl_2$) and mercurous chloride (HgCl) have been used as fungicides. These have been replaced by organic mercury compounds because of their extreme toxicity to plants and long persistence.

## COPPER FUNGICIDES

Copper compounds have wide fungal spectrum and are employed as protective foliar fungicides. Following is list of copper compounds used as fungicides.

***Cupric sulphate:*** $CuSO_45H_2O$ for seed treatement

***Bordeaux mixture:*** many fungal diseases.

$CuSO_4$ (4.57 kg) $Ca(OH)_2$ (5.55 kg) in 454 liters of water $[Cu(OH)_2]_x.CaSO_4$

***Copper dugtdrazubium sulphate:*** Powdery mildew black spot of roses

$$Cu[N_2H_5]_2(SO_4)_2$$

***Copper oxychloride:*** Powdery mildew

$$Cu(OH)_2\ CuCl$$

***Copper zinc chromate:*** diseasese of potato, tomato

$$15CuO.10ZnO.6CrO_3.24H_2O$$

***Cupric carbonate:*** Many fungal diseases

$$Cu(OH)_2\ CuCO_3$$

***Cuprous oxide:*** Powdery mildew

$$Cu_2O$$

***Burgundy mixture:*** Powdery mildew

$$CuSO_4 + Na_2CO_3$$

The mode of action of coper fungicides seem to be inhibitors of the variety of -SH containing enzymes. The fungicidal action of copper continuity fungicides is also associated with the adsorptivity of the cytoplasm of cells of fungi.

## ORGANOTIN COMPOUNDS

### Brestan

(Triphenyl tin acetate)

It is prepared by treating trialkyl tin hydroxide with acetic acid.

$$(C_6H_5)_3\ SnOH + CH_3COOH \longrightarrow (C_6H_5)_3\ SnOCOCH_3 + H_2O$$

Its acute oral $LD_{50}$ for rats is 140-298 mg/kg.

### Duter

(Triphenyl tin hydroxide)

Duter is produced by treating triphenyl stannous chloride with potassium hydroxide.

$$(C_6H_5)_3\ SnCl + KOH \longrightarrow (C_6H_5)\ SnOH + KCl$$

Its acute oral $LD_{50}$ for rats is 171 mg/kg and is used for control of blights of potato, leaf spot on sugar and blast diseases of rice. Complexes such as chlorotriphenyl (Sulfinyl bis (Methane-O)

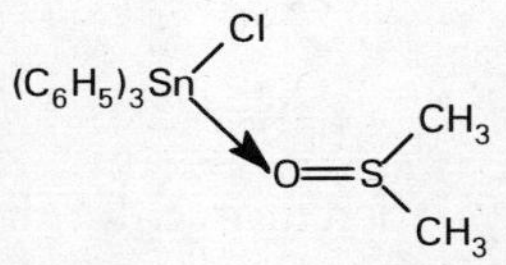


and chlorotriphenyl (quinoline oxide) tin are also fungicides.

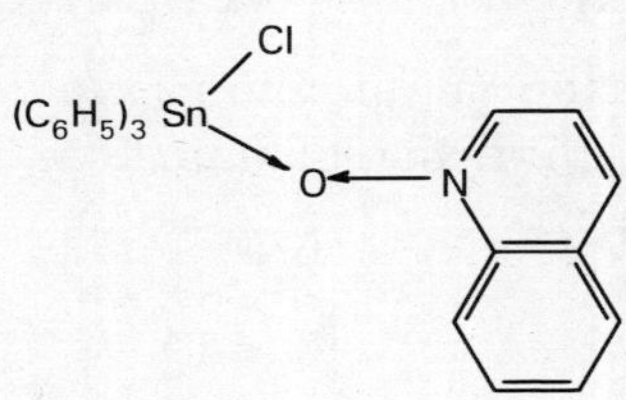


### Mode of Action

The mode of action of tin compounds is due to inhibition of oxidative phosphorylation in the mitochondria. Thus these compounds prevent the incorporation of organic sulphate into highly reactive, high energy ATP or adenosine triphosphate a reaction which typically takes place in mitochondria.

## ORGANOMERCURIALS

Mercuric chloride ($HgCl_2$) is used as bactericide.

Mercurous chloride (HgCl) is much safer for plants than mercuric chloride.

Ethyl mercuric chloride is prepared by treating mercuric chloride and diethyl mercury

$$(C_2H_5)_2\ Hg + HgCl \longrightarrow 2Cl_2H_5HgCl$$

Phenyl mercury acetate is obtainted by reacting benzene with mercuric acetate in glacial acetic acid

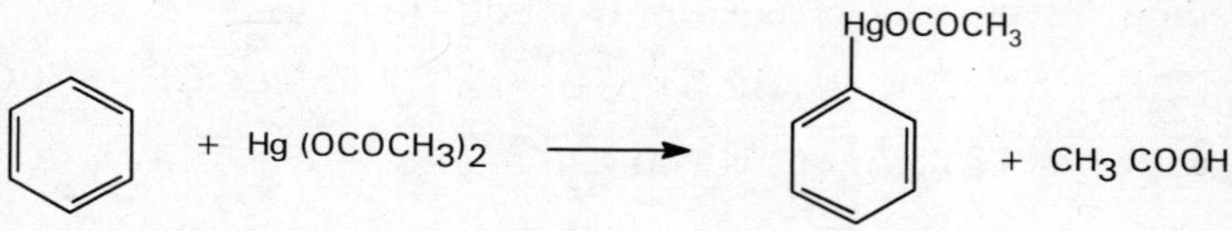


Ceresan-M (N-ethyl mercuric) p-toluene sulphonilide) is produced by reacting ethylmercuric salts with p-toluene sulphonamide in the presence of NaOH

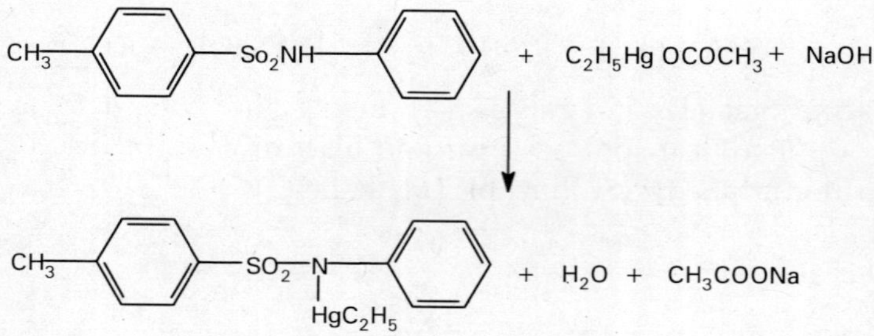


Agallol (methoxy ethyl mercuric chloride) is produced by treating methoxy ethyl mercuric acetate and sodium chloride

$$CH_3OCH_2CH_2HgOCOCH_3 + NaCl \longrightarrow CH_3OCH_2CH_2HgCl + CH_3COONa$$

Phenylmercury triethanol ammonium lactate is produced by treating phenyl mercuric lactate with triethanol amine.

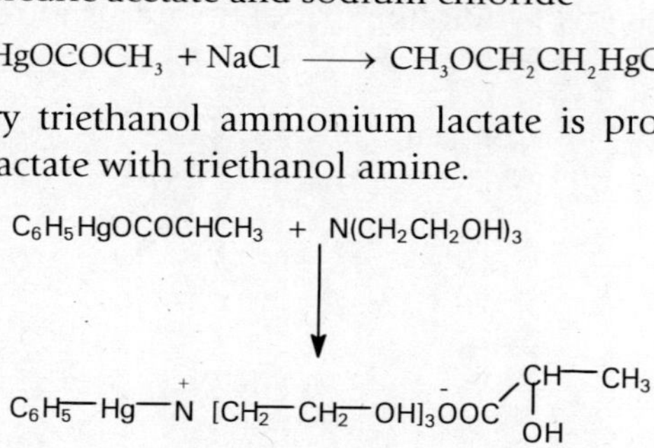

The mercury compounds are used as seed dressing for grains and cereals. The main disadvantage is that they are very poisonous and persistent and there are fears that their continued large scale use may result in serious effects on the ecosystems.

### Mode of Action

The mode of action of the mercury is due to the activity of $Hg^{2+}$ moiety. The biological action of the mercury fungicides is related to the affinity of $Hg^{2+}$ for sulfhydryl groups of essential respiratory enzymes. Thus mercury treatment decreases the oxygen uptake of the treated fungus spores and poisoned the spores.

## ORGANOPHOSPHOROUS COMPOUNDS

### Phosphates

This group has many important systemic fungicides.

### Wepsyn

(P-(5-amino-3-phenyl-1H-1,2,4-triazol 1-yl)-N,N,N'N'-tetra methyl phosphonic diamidate)

Wespyn is prepared by the condensation of sodium salt of 3-amino-5-phenyl-1,2,4 triazole with bis (dimethyl amino) phosphoryl chloride

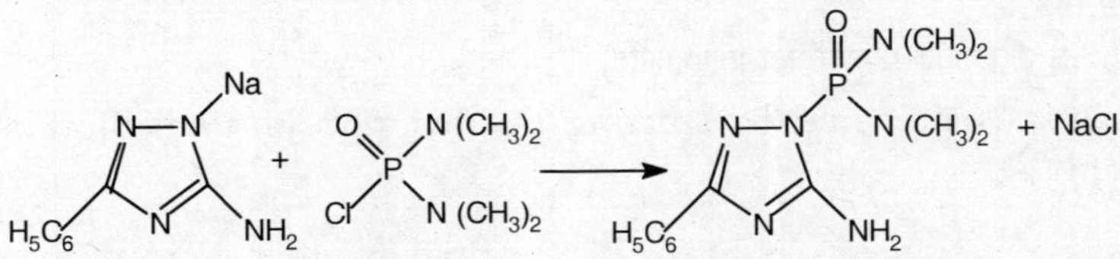

It is used for controlling powdery mildew on apples and ornamentals. The acute oral $LD_{50}$ for rats is 20 mg/kg.

### Kitazin P

(S-Benzyl O,O'-diisopropyl phosphorothiolate)

Kitazin P is prepared by the condensation of sodium benzylthiolate and O,O'-diisopropyl phosphorochloridate.

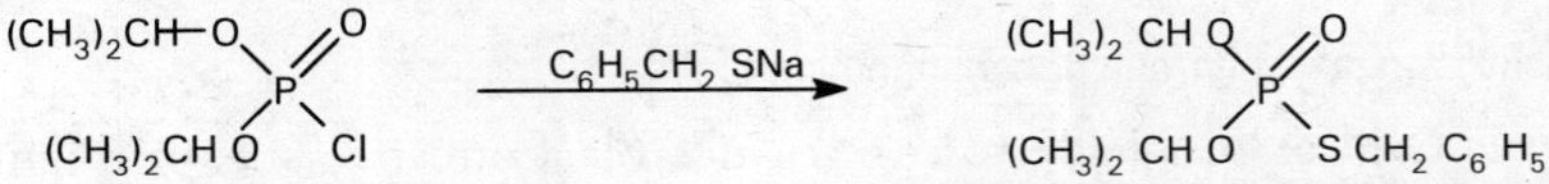

Kitazin P is a systemic fungicide applied as granules to paddy water to control rice blast and it inhibits mycelial growth in the tissue. The acute oral $LD_{50}$ for rats is 660 mg/kg.

### Conen

(O-butyl-S-ethyl-S-benzyl phosphorodithioate)

Conen is prepared from sodium O-butyl-S-benzylphosphoro-dithioate and ethyl bromide.

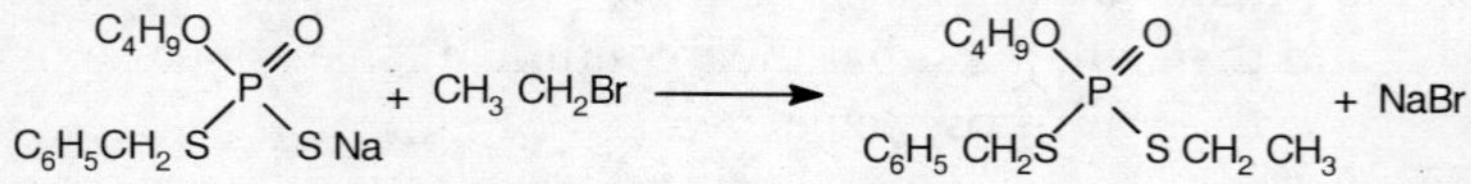

Conen exerts a curative rather than a protective action. The oral $LD_{50}$ for mice is 118 mg/kg.

### Dowco

(O,O-diethylphthalimidophosphorothioate)

This is obtainted by the condensation of potassium phthalimide with O,O-diethyl phos-phorochloridothioate

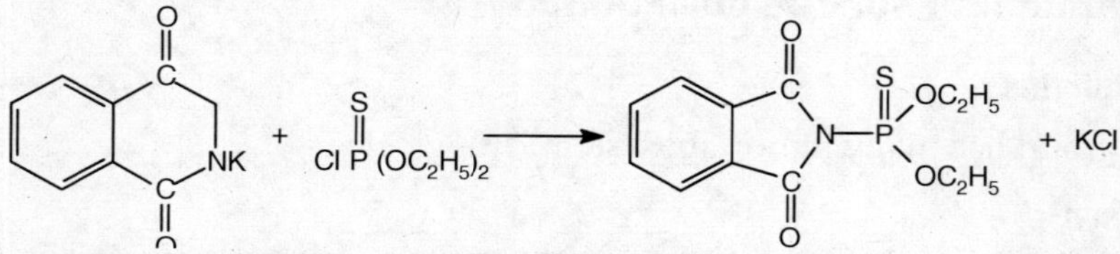

Dowco is very effective against wide range of powdery mildews and has low mammalian toxicity. The oral $LD_{50}$ for rats is 5660 mg/kg.

### Edifenphos

(O-ethyl S,S-diphenyl phosphorodithioate)

Edifenphos is prepared by treating O-ethyl phosphorodichloridate with thiophenol.

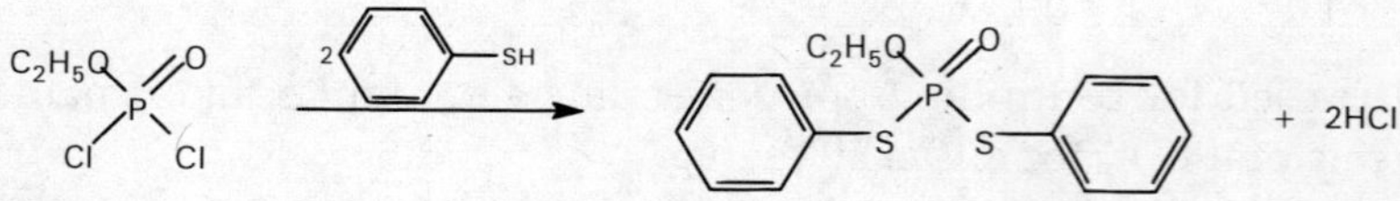

Edifenphos is effective against Pyricularia oryzae on rice. The oral $LD_{50}$ for rats is 218 mg/kg rat.

## PHOSPHONATES

### Inejin F 254

(O-methyl-S-benzyl benzene phosphonothioate)

Inejin is used as fungicide to control rice blast and rice sheath blight. The oral $LD_{50}$ for mice is 720 mg/kg.

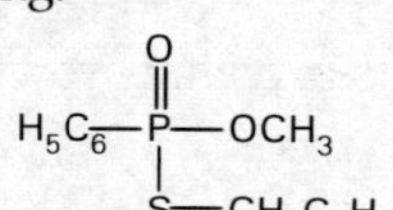

## Cerezin

(O-methyl O-cyclohexyl S (4-chlorophenyl) phosphorothioate)

It is prepared by reacting O-methyl-O-cyclohexylphosphate in carbon tetrachloride with p-chlorophenylsulphenyl chloride

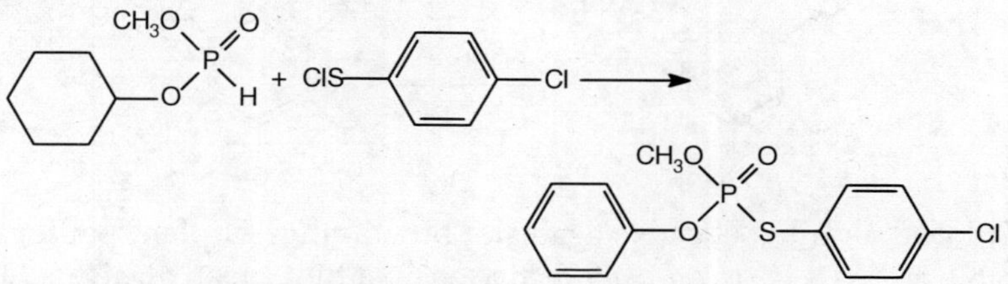

Cerezin has a curative effect on rice blast disease.

## Aliette

(Aluminum tris (O-ethyl phosphonate))

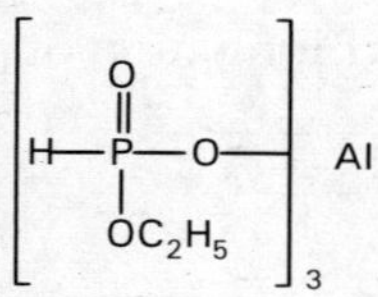

Aliette mainly marketed in combination with a protective fungicide for instance Mikal (50% Aliette, 9.5% Folpel), is active against downy mildews in crops like pineapple, citrus, Avocado, lettuce, vegetables, grapes and ornamentals.

## Mode of Action of Organophosphate Fungicides

The activity of organo phosphorous compounds is due to the reason that these compounds influence fungal metabolism by direct phosphorylation and thereby disturb the vital functions of the cells.

# BENZENE DERIVATIVES

## Quintozene

(2,3,4,5,6-pentachloro-1-nitrobenzene)

Quintozene is prepared by chlorination of nitrobenzene in the presence of chlorosulfonic acid. Iodine is used as catalyst.

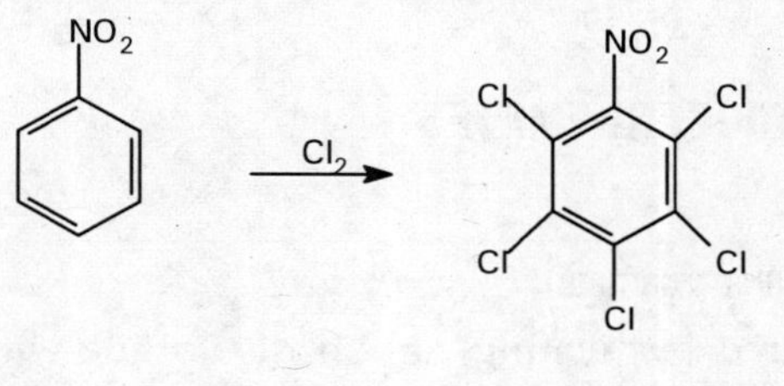

## Chlorothalonil

(2,4,5,6-tetrachloro-isophthalonitrile)

Chlorothalonil is prepared as follows:

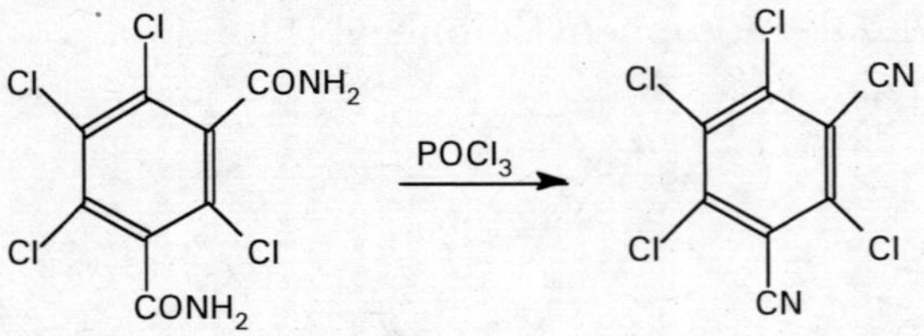

Chlorothalonil is effective against a broad range of plant pathogens on vegetable crops. It is also most effective against blue green algae. Its $LD_{50}$ for rats is 10,000 mg/kg.

## Chloroneb

(1,4-dichloro-2,5-dimethoxybenzene)

Chloroneb is prepared by the chlorination of p-dimethoxybenzene

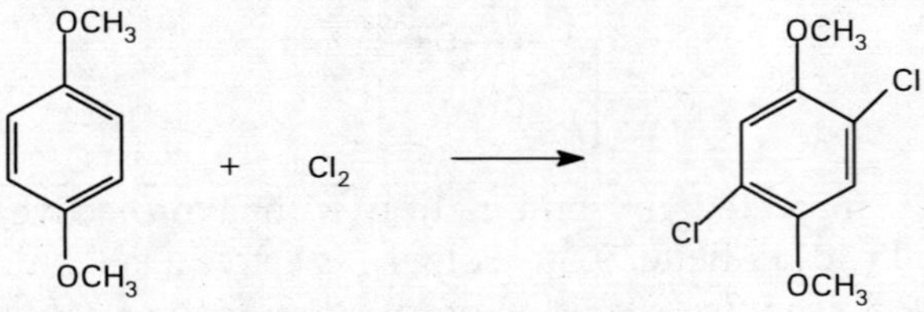

Chloroneb is very efficient for treatment of soil and seeds. It is used for protection of cotton, beans and soybeans. Its acute oral $LD_{50}$ for rats is 11,000 mg/kg.

## DITHIOCARBAMATES

Dithiocarbamates can be divided into two groups

1. N-monoalkyldithiocarbamates
2. N-N'- alkylenedithiocarbamates

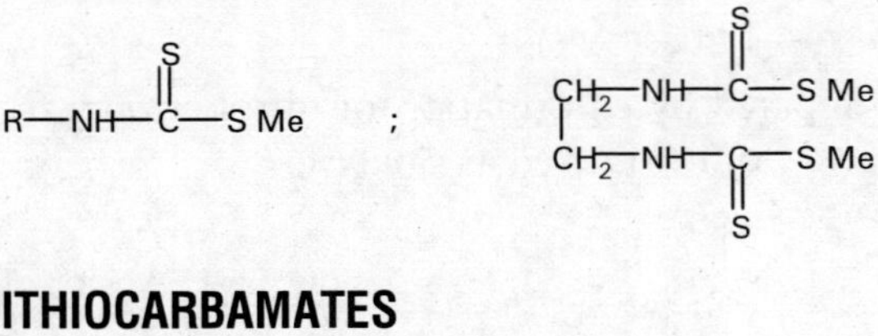

## MONOALKYLDITHIOCARBAMATES

### Metham

(Sodium N-methyldithiocarbamate)

Metham is prepared by treating carbon disulphide with methyl-amine in the presence of sodium hydroxide.

$$CS_2 + CH_3NH_2 \xrightarrow{NaOH} CH_3NH\overset{\overset{\displaystyle S}{\|}}{C}\!-\!SNa + H_2O$$

The fungicidal action of metham is due to liberation of methylisothiocyanate from metham.

$$CH_3NH\overset{\overset{\displaystyle S}{\|}}{C}SH \longrightarrow CH_3\!-\!N\!=\!C\!=\!S + H_2S$$

Metham is used as a seed dressing and soil treatment purposes.

## Dazomet

(3,5-dimethyl-tetrahydro-2H-1,3,5-thiadiazine-2-thione)

Dazomat is prepared by treating methyl amine with carbon disulphide and formaldehyde

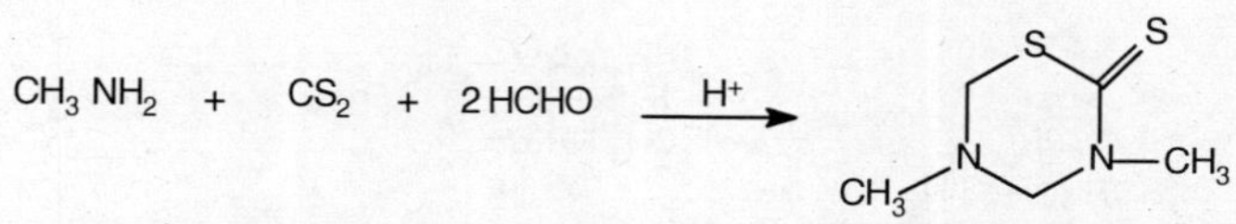

Dazomet on hydrolysis with water liberate formaldehyde and N-methyl dithiocarbamate ester. The later is conveted into methyl isothiocyanate which is responsible for fungicidal action.

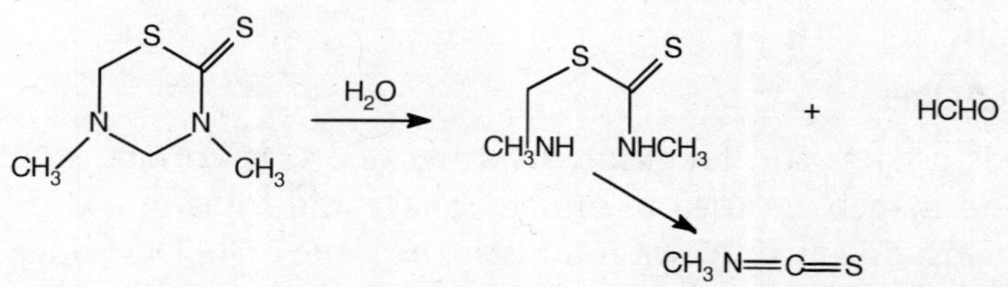

Dazomet is used as soil treatment. It has acute oral $LD_{50}$ of 650 mg/kg to rats.

## Neviram

(Zinc bisisopropyldithiocarbamate)

Neviram is synthesised by treating isopropylamine, carbon disul-phide and ammonium hydroxide. The ammonium isopropyldi-thiocarbamate thus formed, on treatment with zinc sulphate gives zinc bis isopropyl dithiocarbamate. It is used as a foliage fungicide. It can be used against apple and pear scab.

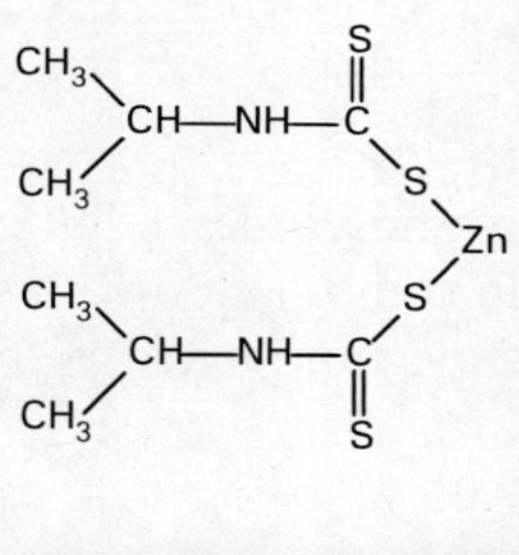

## N,N ETHYLENE-BISDITHIOCARBAMATES

### Nabam

(Disodium ethylene 1,2-bisdithiocarbamate)

Nabam is prepared by reacting sodium hydroxide, carbon disulphide and ethylene diamine in aqueous medium

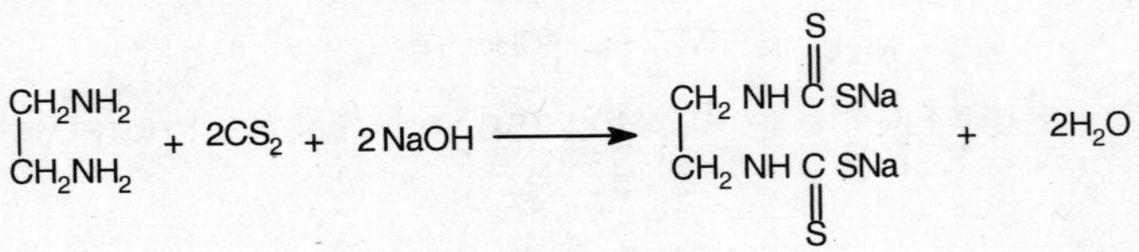

Nabam is converted into either zineb or Maneb by reaction with an aqueous solution of zinc or manganese sulphate.

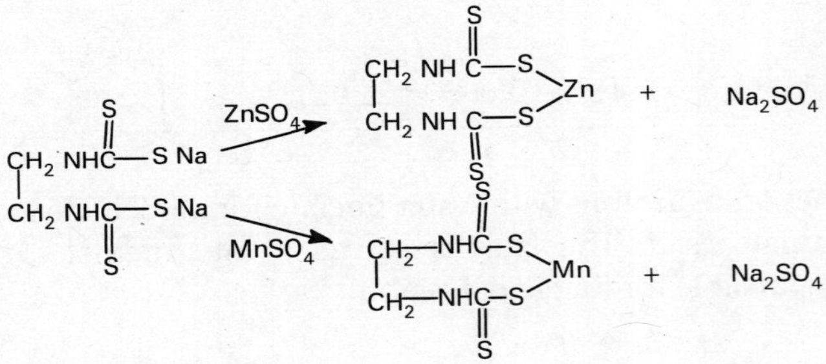

### Mode of Action

Nabam is used as soil fungicide. Its acute oral $LD_{50}$ to rats is 395 mg/kg. Zineb and maneb are used to protect foliage and fruits of a wide range of crops against downy mildews and rusts. The acute oral $LD_{50}$ value to rats for zineb is 5000 mg/kg and for maneb is 6750 mg/kg.

### Mancozeb

(Manzate, Dithane M-45)

Mancozeb is the manganese-zinc double salt of N,N-bisdithio-carbamic acid. The active ingredient is a complex zinc ions (2%); manganese (16%) and ethylene bis dithiocarbamates (62%). Manco-zeb is prepared as follows:

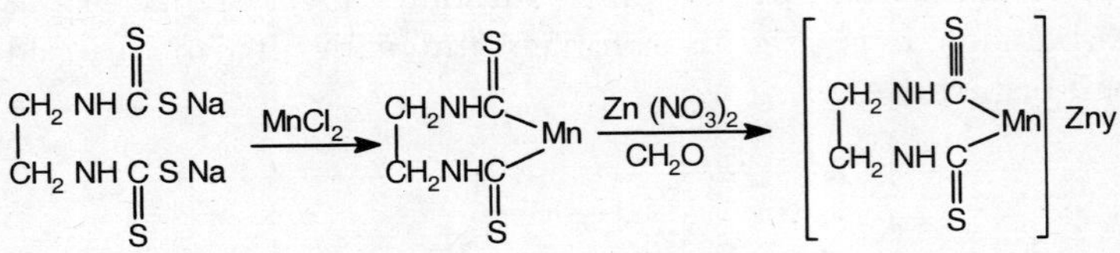

Mancozeb is a protective fungicide and effective against wide range of foliage diseases. It is used in combination with some systemic fungicides to increase the duration of protection given to the crop. Its acute oral $LD_{50}$

value to rats is 8000 mg/kg.

## Mode of Action

The ethylenebisdithiocarbamates act by oxidative decomposition of the fungicides on plant surfaces to diisothiocyanate.

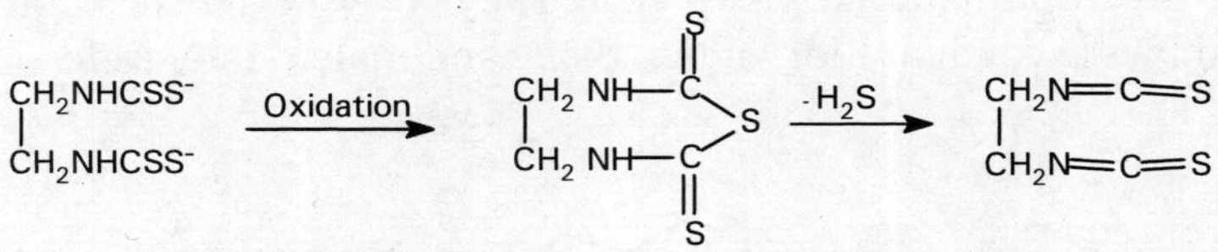


These diisothiocynates are fungicidal by virtue of their ability to react with vital compounds present in the fungal cell.

## Dialkyldithiocarbamates

Ferbam (ferric dimethyl dithiocarbamate), Ziram (zinc bis (dimethyl dithiocarbamate), Thiram (tetramethylthiurum disulphide). These compounds are prepared by the reaction of dimethylamine with carbon disulphide and alkali hydroxide. The individual active substance can be obtainted by the addition of the alkali salt formed.

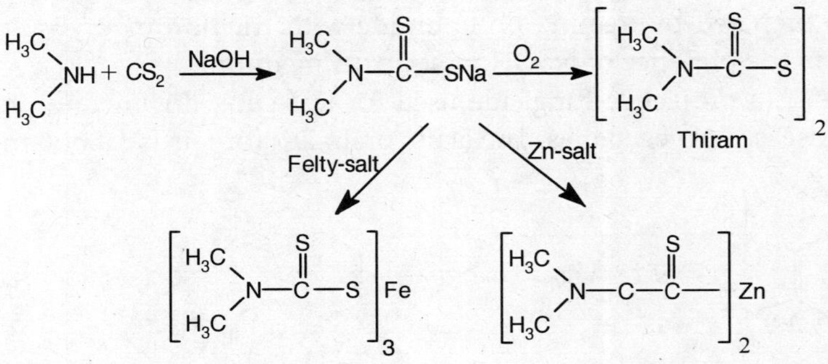


Ferbam Ziram

The Ferbam, ziram and thiram are effective against apple scab and blue mold and are also effective against the seedling diseases of vegetables. The mode of action of these dialkyl dithiocarbamates is due to their ability to chelate with metals especially copper.

# COMPOUNDS CONTAINING POLYHALOGEN ALKONOIC SULFENYL GROUP

## Captan

(N-(trichloromethylthio)-cyclohex-4-ene-1,2-dicarboximide)

Captan is a heterocyclic nitrogenous compound. It is produced by the condensation of trichloromethylsulphenyl chloride with tetrahydrophthalimide in aqueous alkaline medium.

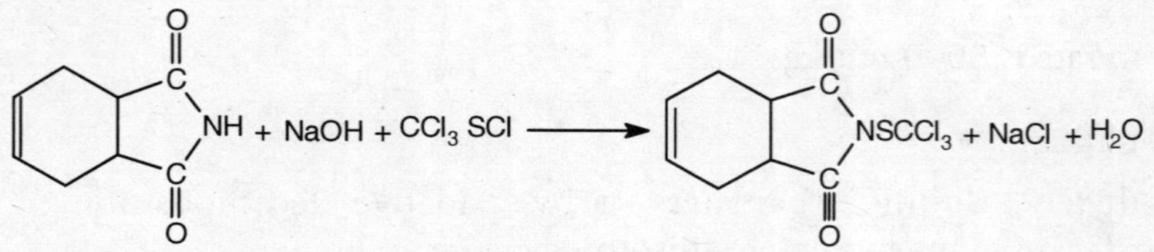

Tetra hydrophthalimide is prepared by Diels-Alder reaction. It is obtainted by the condensation of butadiene and maleic anhydride

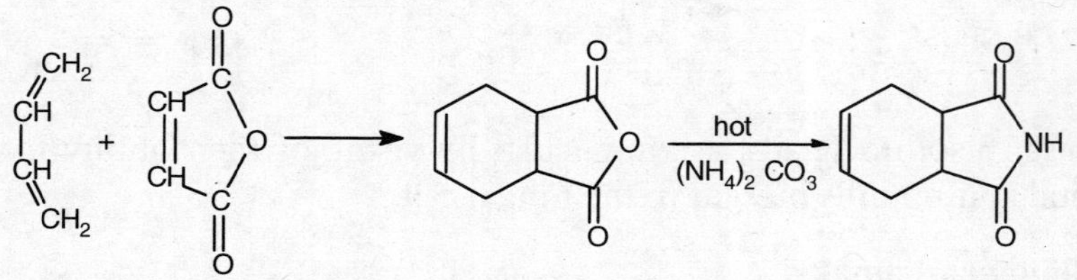

Captan is a fungicide with a very wide range. Captan is foliage fungicide with protective action but is manly used for seed treatment and soil fungicide. The acute oral $LD_{50}$ for rats is 900 mg/kg.

## Folpet

(N-(trichloromethyl sulphenyl phthalimide)

Folpet is prepared by treating phthalimide with trichloromethyl-sulphenyl chloride in the presence of NaOH in aqueous medium.

Folpet is a protective fungicide used for leaf spot, downy and powdery mildew diseases of vegetables. The acute oral $LD_{50}$ for rats is 10,000 mg/kg.

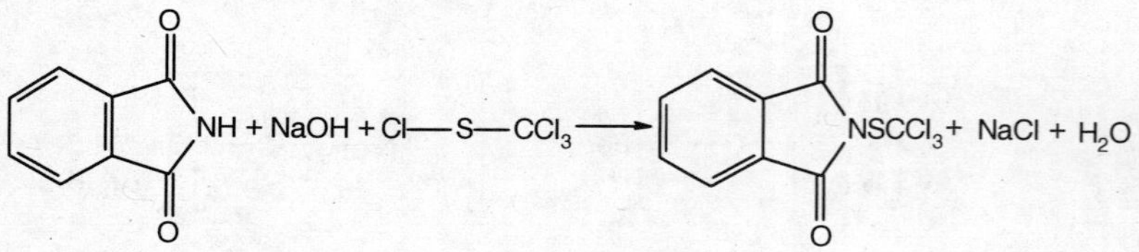

## Captafol

(N-((1,1,2,2-tetrachloroethyl) thio) 4-cyclohexene-1,2-dicarboximide)

Captafol is produced by the reaction of tetrahydrophthalimide with pentachloroethylmercaptan in the presence of NaOH

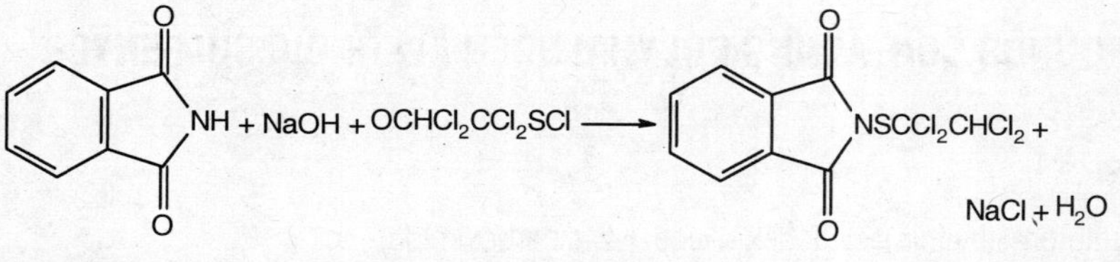

Captafol is protective, wide spectrum foliage and soil fungicide. Its acute oral $LD_{50}$ for rats is 6000 mg/kg.

### Mesulfan

(N-trichloromethylthio-4-chloroanilide of methane sulphonic acid)

Mesulfan is produced by reacting trichloromethylsulphenyl chloride with 4-chlorophenylmethane sulphonimide in the presence of NaOH

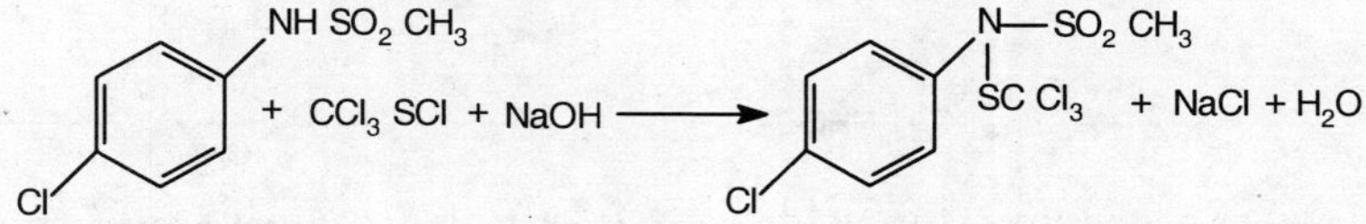


### Dichlofluanid

(N'-dichlorofluoromethylthio-N,N-dimethyl-N'-phenylsulfamide)

Dichlofluanid is prepared by treating sulfuryl chloride with dimethylamine and aniline in turn. N,N-dimethyl-N'- phenyl-sulfamide which is formed as an intermediate product is then coupled with fluorodichloromethanesulfenyl chloride to yield dichlofluanide.

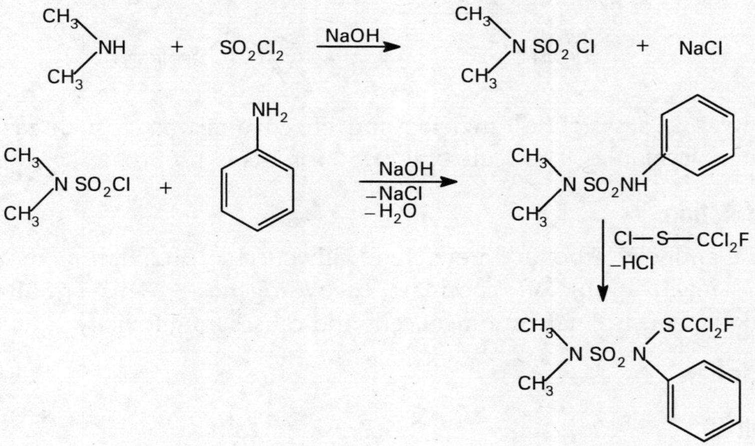


## OXATHIINS

### Vitavax

(carboxin, 2,3-dihydro-5-carboxanilid-6-methyl-1,4 oxathiin)

Vitavax is prepared by the reaction acetoacetanilide with sulfuryl chloride to give chloroaceto acetanilide. This on reaction with 2-thioethanol followed by cyclization give vitavex.

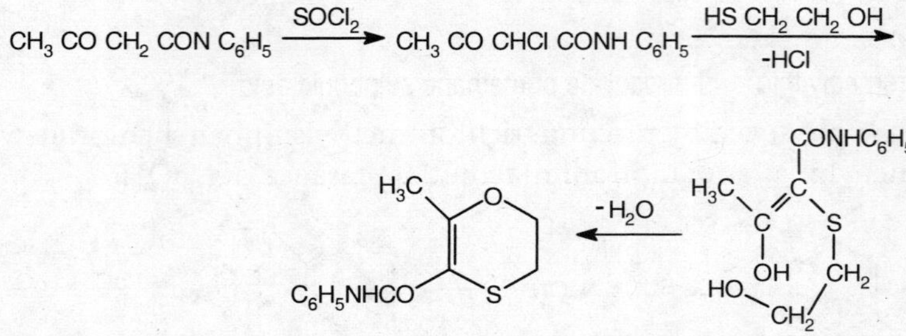


Vitavax is a systemic fungicide and is effective against loose smut of barley and wheat. The acute oral $LD_{50}$ for rats is 3820 mg/kg.

## Plantvax

(Oxycarboxin, 2,3-dihydro-5-carboxanido-6-methyl 1,4 oxathiin-4,4- dioxide)

It is prepared by the oxidation of carboxin with hydrogen peroxide.

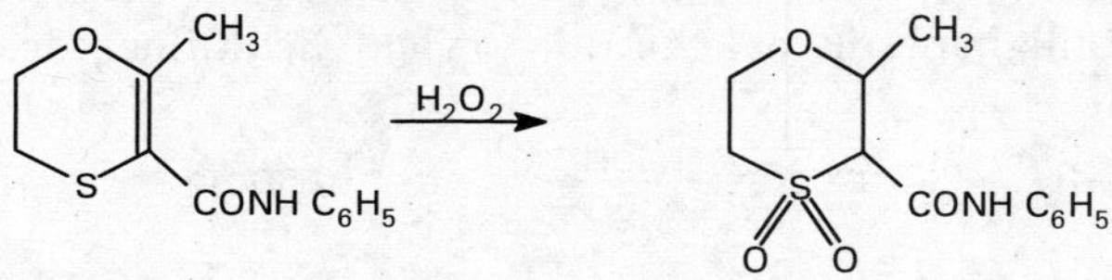


Plantvax is a systemic fungicide and effective against rust diseases of cereals and vegetables. The acute oral LD50 for rats is 12000 mg/kg.

## Mode of Action

Carboxin and oxycarboxin interefered with succinate oxidation in fungi through inactivation of succinate dehydrogenase. This results in accumulation of succinate in fungal cells and causes fungitoxicity.

# QUINONES

## Chloranil

(Tetrachloro-p-benzoquinone)

It is prepared by oxidative chlorination of phenol or by oxidative hydrolysis of pentachlorophenol

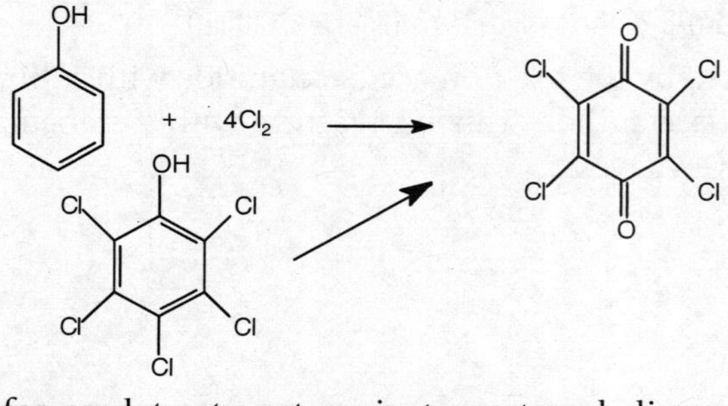


It is used for seed treatment against smut and diseases of vegetable

seeds. The acute oral $LD_{50}$ for rats is 4000 mg/kg.

## Dichlone

(2,3-dichloro-1,4-napthaquinone)

Dichlone is produced by bubbling chlorine gas through a solution of 1,4-aminonaphthalene sulphonic acid in aqueous sulphuric acid.

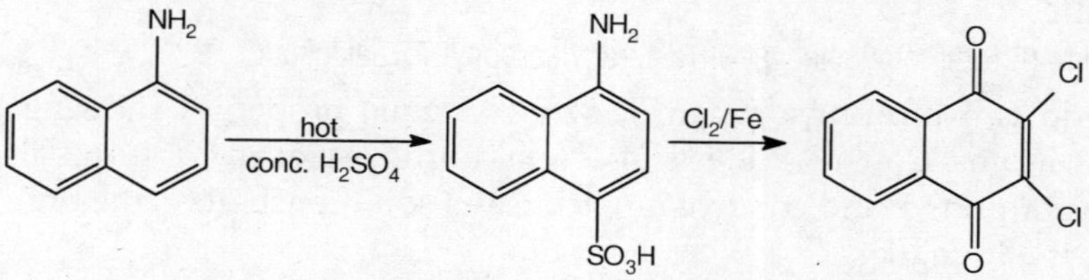


It can be used as foliage fungicide against powdery mildew of apple scab and gay mould.

## Dithianon

(Delon, 5,10-dihydro-5,10-dioxonaphthol [2,3-b] 1,4 dithiin-2,3-dicarbonitrile)

It is prepared by the reaction of dichlone with 2,3-mercapto 2-butenedinitrile

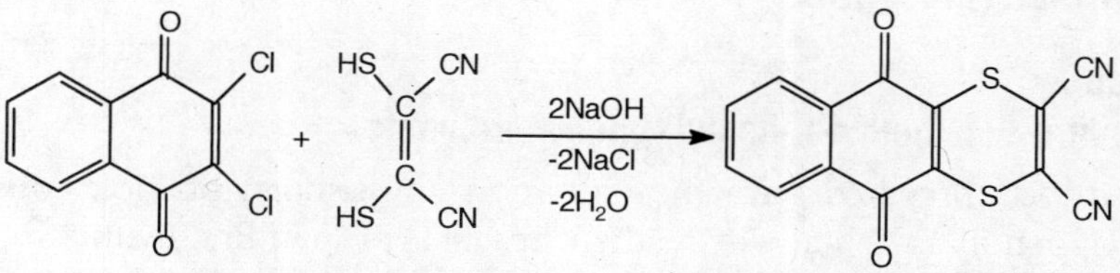


Dithianon is a protective fungicide used mainly against apple and pear scab.

## Mode of Action

The fungicidal properties of quinones are due to an addition reaction with sulphydryl containing respiratory enzymes in the fungal cell. The overall toxic mechanism therefore probably consists in binding the enzyme to the quinone nucleus by substitution or addition to the double bond.

# CARBOXAMIDES

## Metalaxyl

(Methyl-N-(2,6-dimethyl phenyl) N-(2-methoxyacetyl) DL-alaninate)

It is effective against species causing downy mildews, late blight, damping off root, stem and fruit roots.

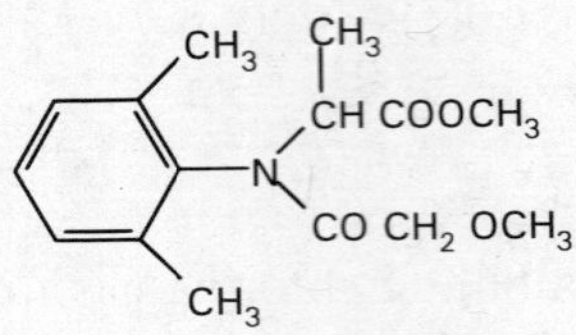


**Fongarid**

(Methyl N-(2,6-dimethylphenyl)-N- (2-furanylcarbonyl) DL-alaninate)

Fongarid is a protective fungicide with systemic properties showing high activity against downy mildews of several horticultural crops. It can also be used as protective and curative fungicide as a soil drench. Its acute oral $LD_{50}$ to rats is 940 mg/kg.

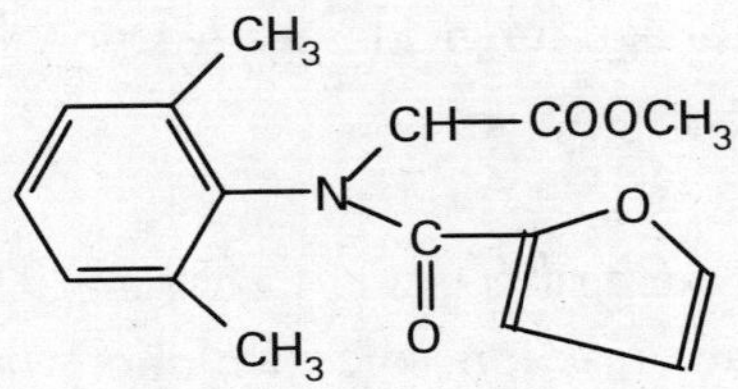


## DINITROALKYLPHENOLS

### Dinocap
### (2,4-dinitro-6(1-methyl-n-heptyl) phenyl crotonate

Dinocap is prepared by the nitration of isooctylphenol, dinitro-isooctylphenol so obtained is then reacted in pyridine with crotonyl chloride

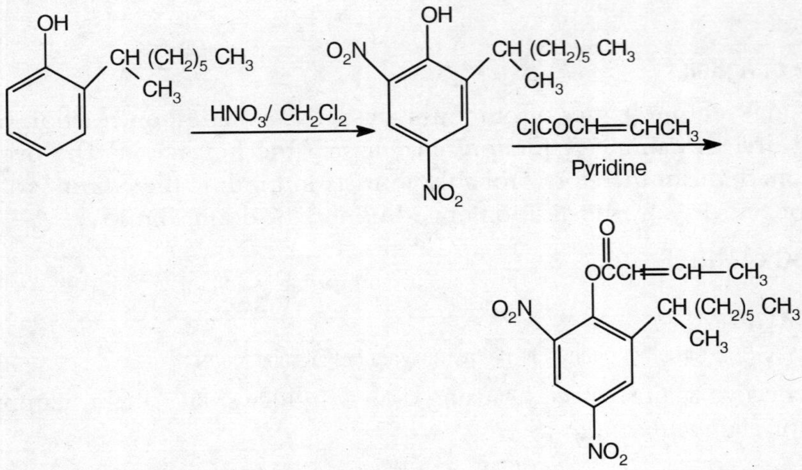


Dinocap is effective only against powdery mildew fungi. It is used mainly in orchards, vine cultures, vegetables and for protection of ornamental crops. Its acute oral $LD_{50}$ to rats is 980 mg/kg.

### Binapacryl

(4,6-dinitro-2-sec-butyl-phenyl-3-methyl crotonate)

Binapacryl is prepared by esterification of dinoseb with 3-methyl crotonic acid chloride

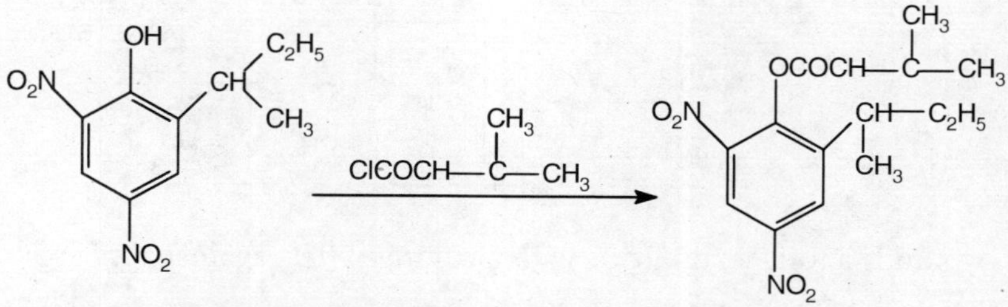

It is mainly used against powdery mildew in orchards.

### Dinobuton

(2-sec-butyl-4,6-dinitrophenyl isopropyl carbonate)

Dinobuton is prepared by treating alkali salt of dinoseb with isopropyl chloroformate

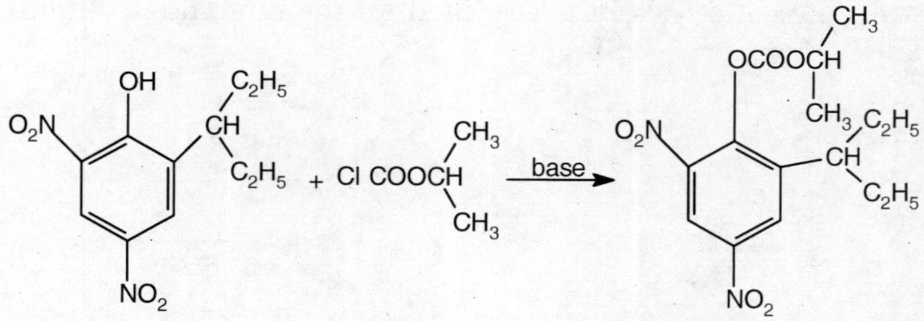

Dinobuton is a contact fungicide and is effective against powdery mildew and is used on tomatoes.

## AZOLE COMPOUNDS

### Imidazole and its Derivatives

### Glyodin

(2-heptadecyl-2-imidazoline acetate)

Glyodin is prepared by treating ethylene diamine and ethyl stearate, forming its acidic amide which on heating is cyclised and forms a salt with acetic acid.

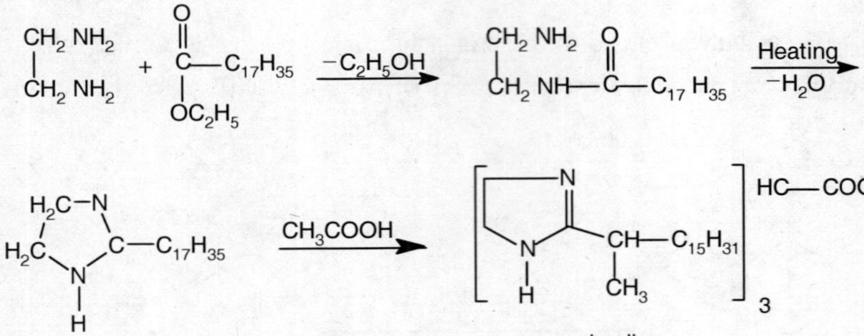

glyodin

Glyodin is a foliage fungicide with a curative effect against apple scab.

### Imazalil

(1 -ß- propenyloxy-2,4-dichlorophenyl imidazole)

Imazalil can be prepared by reduction of intermediate product formed in the reaction of chlorophenacyl bromide with imidazole and subsequent alkylation with allyl chloride

Imazalil is a fungicide with systemic action. It is ued as seed dresser against diseases of cereals. It is also used for the protection of fruits and vegetables.

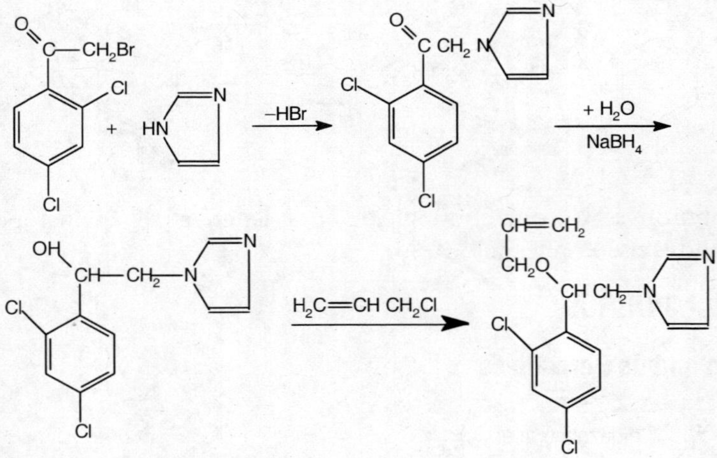

## BENZIMIDAZOLE DERIVATIVES

### Carbendazim

(Methyl-N-benzimidazol-2-yl carbamate)

Carbendazim is prepared by methylation of thiourea with dimethyl sulphate and subsequent N-acylation in alkaline medium of the S-methyl-isothiourea formed with methyl ester of chloroformic acid. The N,N'-bis (methoxycarbonyl)-S-methyl isothiourea is condensed with O-phenylene diamine in glacial acetic acid.

Carbendazim is a systemic fungicide controlling a wide range of pathogens of cereals, fruits, grapes, ornamentals and vegetables. It is very efficient against leaf and ear diseases of wheat but it is inactive against rust. The acute oral $LD_{50}$ for rats is 15000 mg/kg.

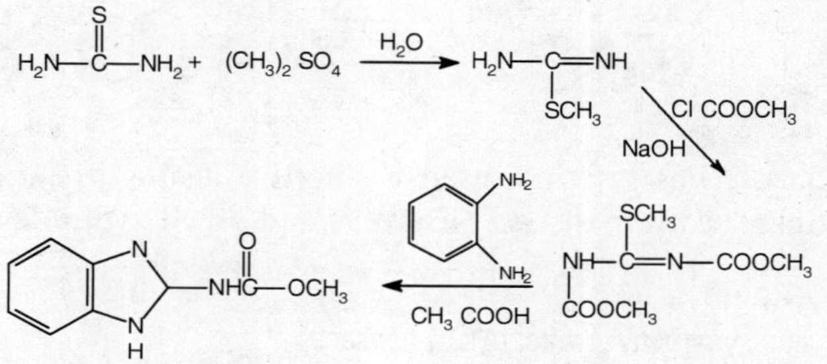

## Benomyl

(N-(1-butylcarbamoyl-benzimidazole 2-yl) methylcarbamate)

It is produced by the reaction of methyl-N-benzimidazol-2-yl-carbamate with butyl isocyanate.

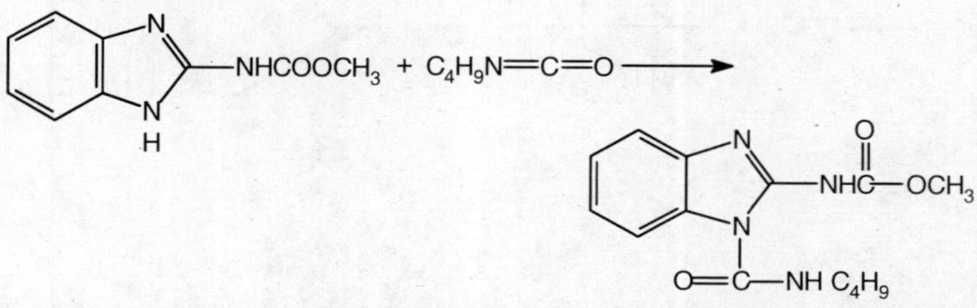

Benomyl has a protective, and eradicant fungicide with systemic activity. Benomyl is absorbed by plants and rapidly metabolized into MBC. Benomyl is very effective against a wide range of fungi affecting fruits, nuts and vegetables. It is also a very good seed protectant. The acute oral $LD_{50}$ for rats is 10,000 mg/kg.

## Thiabendazole

(2-(4'-thiazolyl)-benzimidazole)

It is prepared from O-phenylenediamine and 4-thiazole carboxamide in polyphosphoric acid. Thiabendazole is used in the protection of apples and pears against fungi during storage. It is also very effective as seed treatment against bunt of wheat.

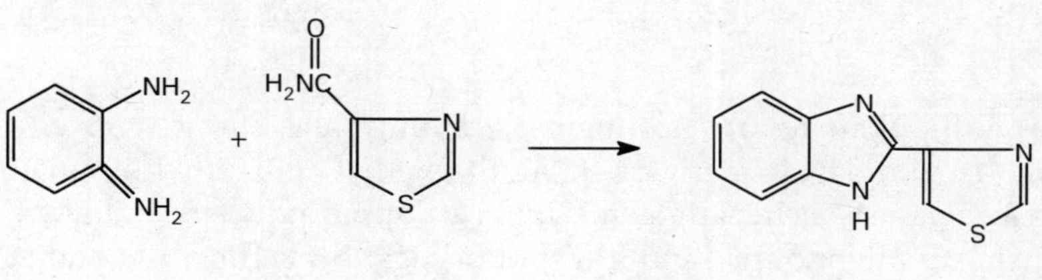

## Fuberidazole
## ((2,2'-furyl) benzimidazole)

It is prepared by the reaction of furfural with O-phenylenediamine in the presence of oxidising agent cupric acetate in the medium of methyl alcohol.

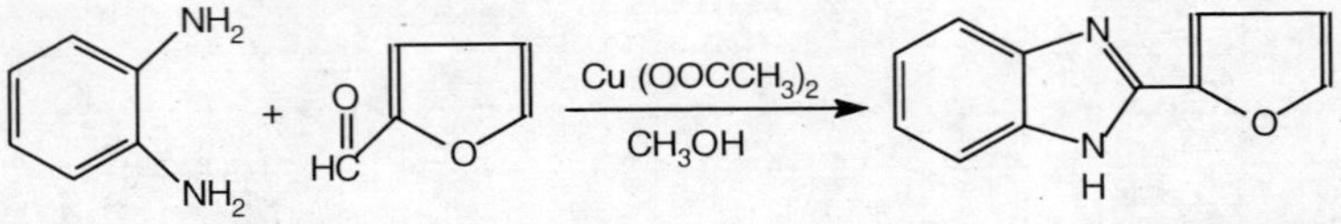


Fuberidazole has systemic used is effective against treatment of seeds mainly against Fusarium diseases leaf rust, and powdery mildew on barley.

## Thiophanate-methyl

(1,2-bis (3-methoxycarbonyl 1-thiourelido) benzene)

Thiophanate methyl is prepared by the reaction of sodium or potassium thiocyanate with chloroformic acid methyl ester and the intermediate product reacts with O-phenylenediamine to form thiophanate methyl

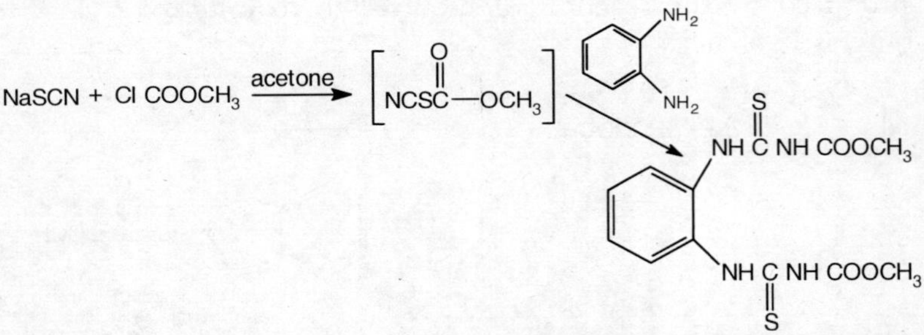


Its ethyl homologue is thiophanate and is commonly known as tapsin. It is prepared by condensation of sodium thiocyanate, ethylchloroformate and phenylenediamine.

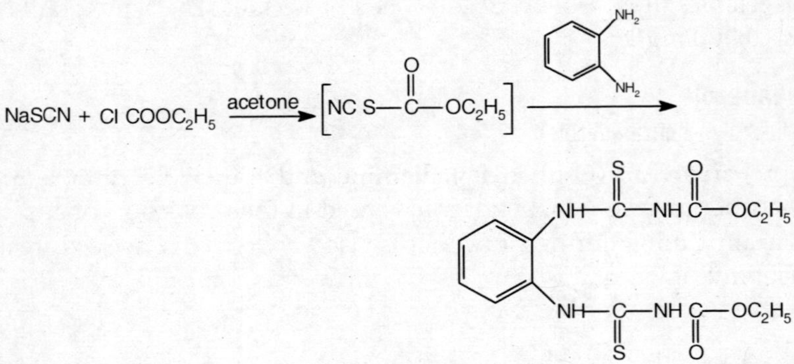


The thiophanates are not themselves fungicidal but are converted into active benzimidazole derivatives. Both methyl and ethylthiophante are effective against such pathogenic fungi as apple powdery mildews, apple scab, sheath blight of rice and leaf spot of sugar beet. The acute oral $LD_{50}$ for

thiophanate ethyl for rats is 15000 mg/kg, for thiophanate methyl it is 2500 mg/kg.

## TRIAZOLES

### Fluotrimazole

(1-(3-trifluoromethyltrityl)-1H-1,2,4-triazole)

Fluotrimazole is prepared by treating m-trifluoromethylphenyl bromide with benzophenone in grignard synthesis. The resulting 3-trifluoromethylphenyl-bis (phenyl) methylcarbinol obtainted is converted to tritylchloride and is treated with triazole in acetone solution to yield fluotrimazole.

Fluotrimazole is effective against all powdery mildew fungi on cereals, apples and roses. The acute oral $LD_{50}$ for rats is higher than 5000 mg/kg.

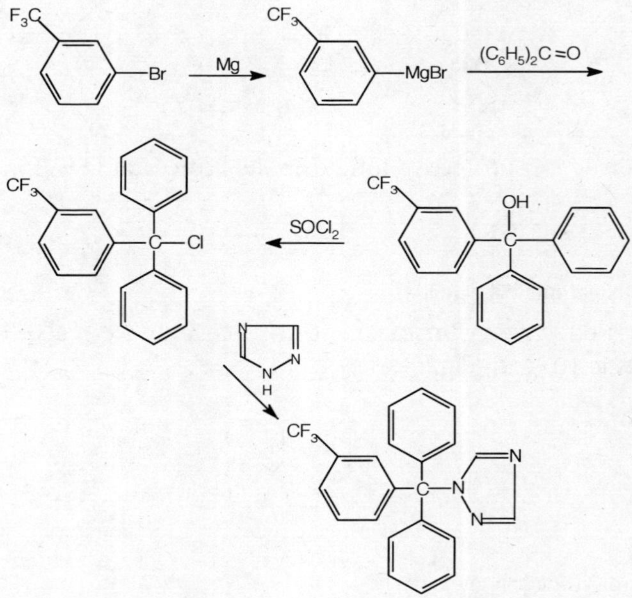

### Triadimefon

(1-(4-chlorophenoxy)-3,3-dimethyl-1-(1,2,4-triazole- 1-yl) butan- 2-one)

Triadimefon is a fungicide with excellent broad spectrum of action. It is effective against mildews and rusts attacking cereals, coffee, grapes and vegetables. Its acute oral $LD_{50}$ for rats is 5 to 8 mg/kg.

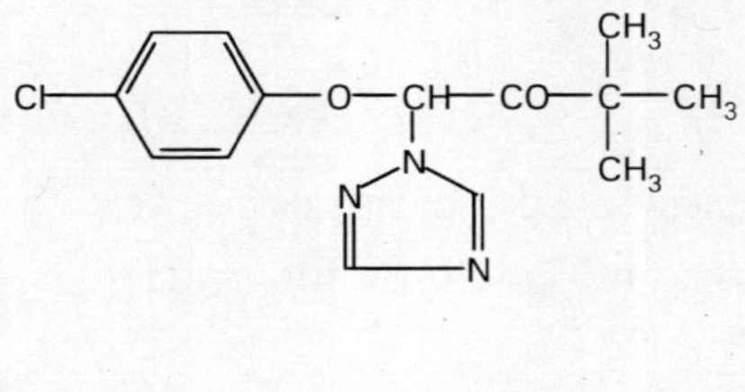

# HETEROCYCLIC FUNGICIDES

## Pyridine Derivatives

### Pyridinitrile

### (2,6-dichloro-4-phenylpyridine-3,5-dicarbonitrile)

Pyridinitrile is prepared by the reaction of cyanoacetic acid ethyl ester with ammonia followed by coupling of two molecules of cyanoacetic amide formed with benzaldehyde and by chlorination of the product, 3,5-dicyano-4-phenyl-6-hydroxyhydro-2-pyridine

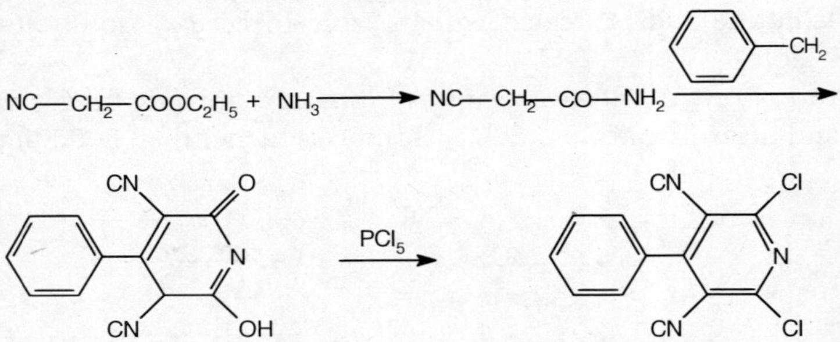

Pyridinitrile is a protective fungicide. Its acute oral $LD_{50}$ for rats is 5000 mg/kg.

### Nitrapyrin

(2-chloro-6-trichloromethyl pyridine)

Nitrapyrin is effective against stem rot of maize and cereals. Its acute oral $LD_{50}$ for rats is 1072 mg/kg.

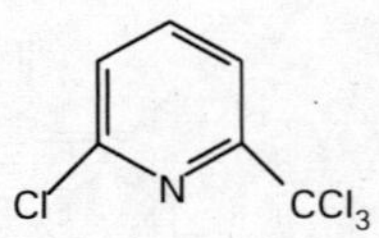

### Dowco 263

(3,4,5-trichloro-2,6-dicyanopyridine)

Dowco-263 is found to be effective for seed treatment against soil fungi and in cereals and cotton. Its acute oral $LD_{50}$ for rats is 562 mg/kg.

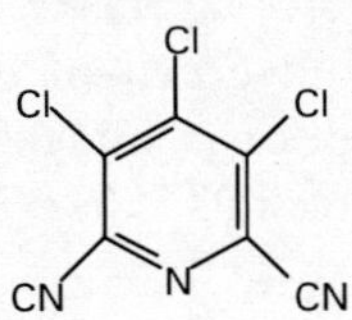

### Piperalin

(3-(2-methylpiperidino) propyl 3,4-dichlorobenzoate)

Piperalin is protective and eradicant fungicide It is specific fungicide for ornamental plants against powdery mildew fungi. Its acute oral $LD_{50}$ for rats is 2500 mg/kg.

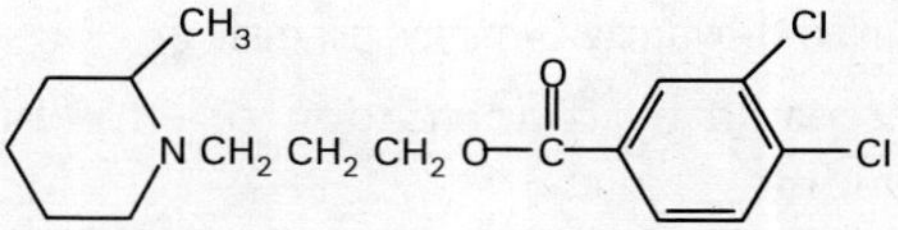

### Parinol

(3-(hydroxy-bis (4-chlorophenyl) methyl) pyridine)

It is effective against the powdery mildew fungi of ornamental plants. Its oral $LD_{50}$ for rats is 5000 mg/kg.

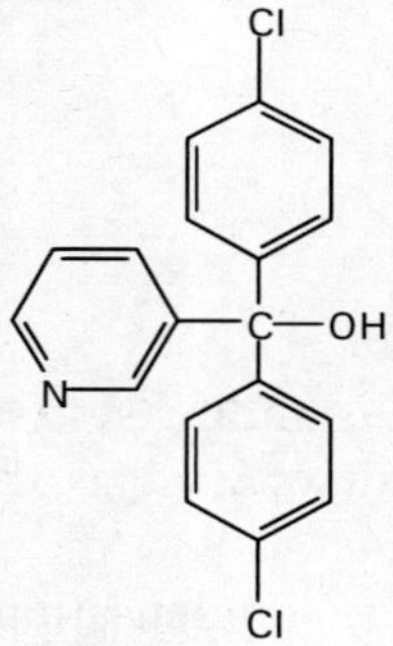

## PIPERAZINE DERIVATIVE

### Triforine

(N,N'-(1,4-piperazinediyl-bis (2,2,2- trichloroethylidene)) bis formamide)

Triforine is prepared by the reaction of piperazine with 1,2,2,2 - tetrachloroethyl formamide.

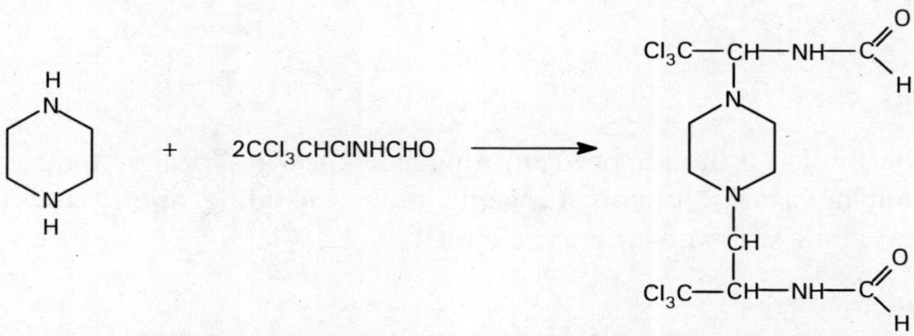

Triforine is systemic fungicide effective for the control of powdery mildew, scab, rust, and leaf spot diseases on a wide range of crops.

## PYRIMIDINES

### Dimethirimol

(5n-butyl-2-dimethylamino-4-hydroxy-6-methyl pyrimidine)

Dimethirimol is prepared by condensation of ethyl butylaetoacetate with N,N-dimethyl guanidine.

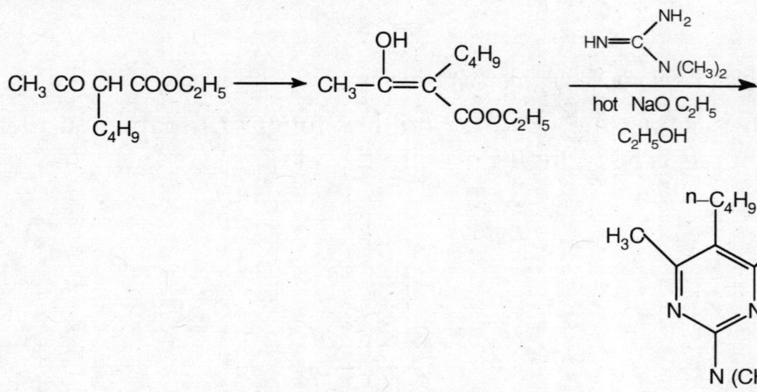

Dimethirimol showed outstanding systemic activity by root application against certain powdery mildews such as those of cucumber, melon and some ornamentals.

**Triarimol** is obtainted by condensation of 5-bromopyrimidine with 2,4-dichlorobenzophenone.

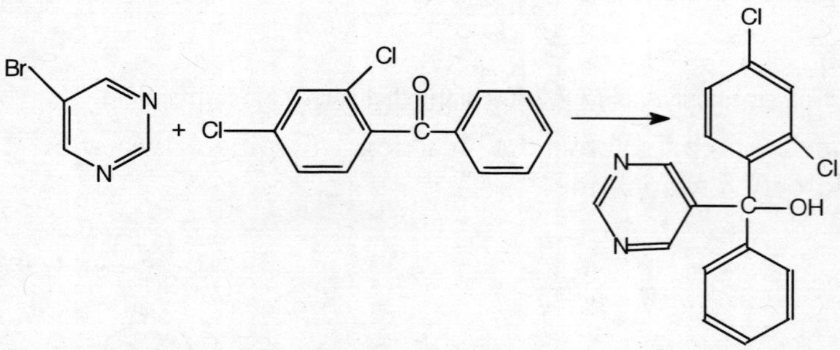

Triarimol is a broad spectrum fungicide showing local systemic action by spraying against phytopathogenic fungi including apple, and barley mildew, apple scab and cherry leaf spot.

### Ethirimol

(5-n-butyl-2- ethylamine-4-hydroxy-6- methylpyrimidine)

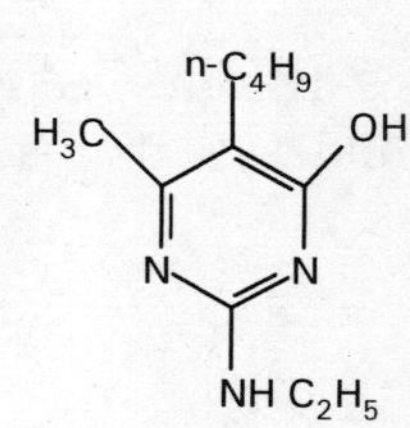

Ethirimol is toxic to powdery mildew fungi acting specifically against this parasite. It is also effective against powdery mildew of tobacco, peas and hops, Its acute oral $LD_{50}$ to rats is:

## Bupirimate

(5-n-butyl-2-ethylamino-6-methyl pyrimidine-4-yl dimethylsulphamate)

Bupirimate is a foliage fungicide. It is effective against apple powdery mildew, cucurbit and rose powdery mildews. Its oral $LD_{50}$ for rats is 4000 mg/kg

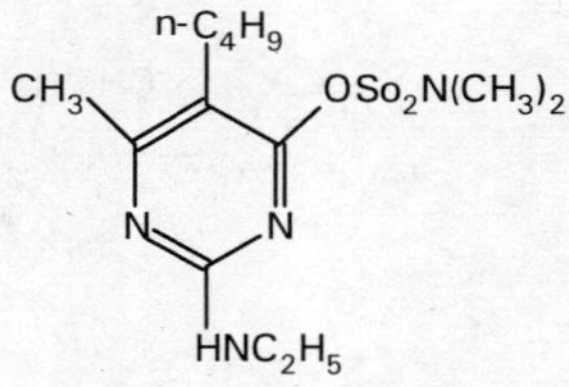


## Fenarimol

(2,4'-dichloro-pyrimidin-5 yl) benzhydryl alcohol)

Fenarimol is prepared by treating 5-bromopyrimidine with 2,4'-dichloro benzophenone

Fenarimol is having a protective, curvative, eradicative activity against powdery mildews and recommended for use on apples, cucurbits, grapes and on roses. It also demonstrates activity against apple scab and against leaf spot diseases of peanut, soybean and sugarbeet. Its $LD_{50}$ for rats is 2500 mg/kg.

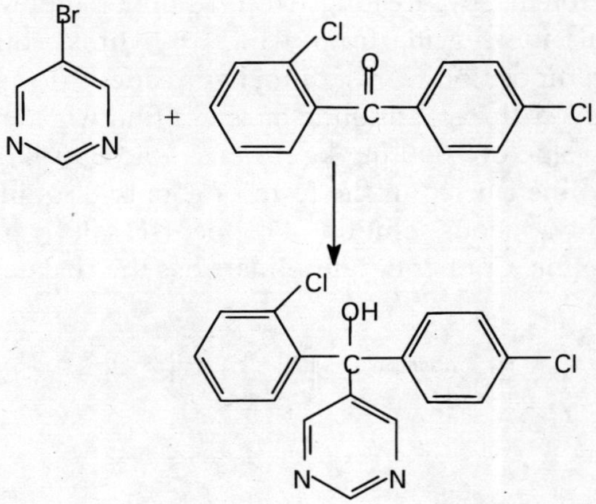


## Nuarimol

((2-chlorophenyl) (4-fluorophenyl)-5- pyrimidinemethanol)

Nuarimol is synthesised by the reaction of 5-bromopyrimidine with 2-chloro-4'-fluorobenzophenone.

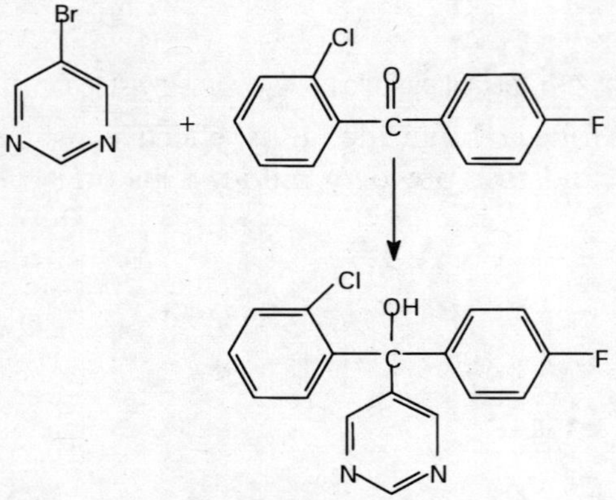

Nuarimol has demonstrated broad spectrum of antifungal activity. As a foliar application, it has provided control of powdery mildews on apple, cherry, cucurbits, grapes, wheat and of leaf spot diseases of strawberry, cherry, soybean. The acute oral $LD_{50}$ for mice and rats are 2700 and 1600 mg/kg respectively.

## QUINOLINE DERIVATIVES

### 8-Hydroxyquinoline

8- Hydroxyquinoline is obtainted from quinoline, which is formed from aniline and acrolein, by treating with fuming sulphuric acid, giving quinoline -8-sulphonic acid, from which 8-hydroxy quinoline can be prepared by decomposition with sodium hydroxide.

As fungicide 8-hydroxy quinoline does not find wide application. It has acute oral $LD_{50}$ value of 1000 mg/kg for rats. For control of plant diseases, 8-hydroxy quinoline is used in the form of copper-8- quinolinate which is prepared from the aqueous solution of copper (II) salt by precipitation with 8-hydroxy quinoline. Copper-8- quinolinate has the chelate structure.

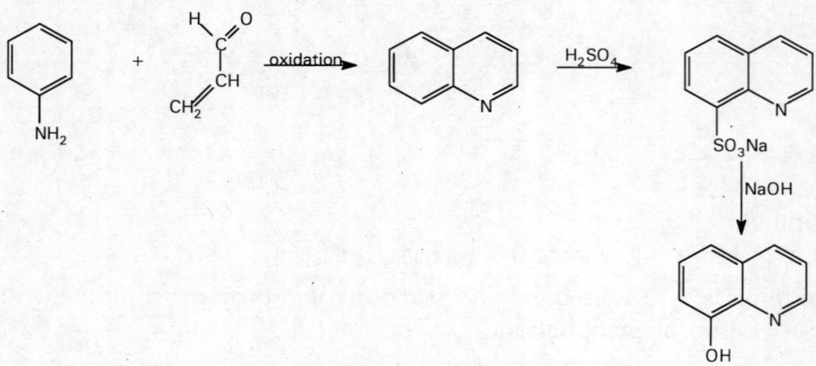

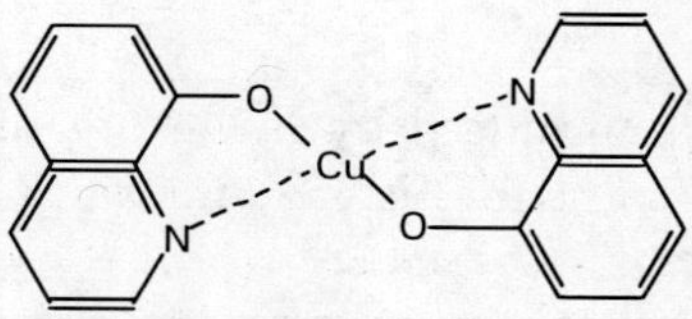

### Quinacetol Sulphate

(bis (5-acetyl-8-hydroxyquinolinium) sulphate)

Title compound is a derivative of 8-hydroxyquinoline and is recommended for seed treatment of wheat, barley and oats. Its acute oral $LD_{50}$ for rats is 2200 mg/kg.

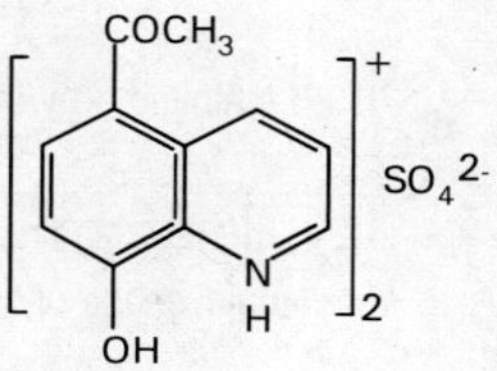

## MORPHOLINE TYPE FUNGICIDES

### Morpholine

Morpholine can be obtainted in a satisfactory yield by the cyclisation of bis (2-hydroxyethyl) amine with the elimination of water.

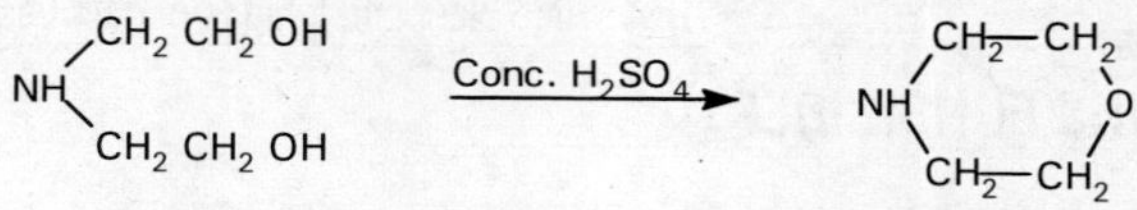

Two morphine fungicides tridemorph (2,6-dimethyl-4-tridecylmorpholine) and cyclododecyl derivative known as dodemorph (4-cyclododecyl 2,6-dimethyl morpholine) are active against powdery mildews.

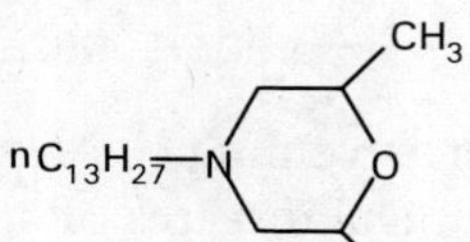

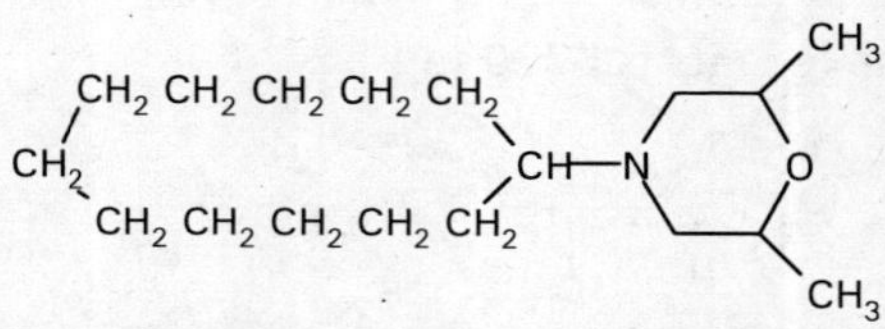

Morpholine type fungicides have an eradicant action when applied on leaves.

## ANTIBIOTICS

Antibiotics are chemicals produced by living organisms, that are toxic to other organisms. Following is the list of antibiotics, their structure, source and uses as the fungicides.

| Name | Source | Structure and Uses |
|---|---|---|
| Gliotoxin | Trichoderma viride | 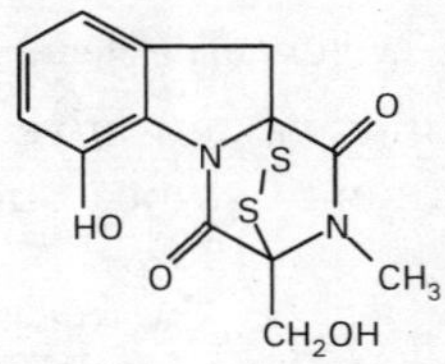 **USE:** Inhibited growth of Botrytis and Fusarium spores |
| Streptomycin | Streptomyces griseus | **USE:** Effective against bacterial pathogens of plants and bacterial disease of stone fruits |
| Blasticidin | Streptomyces griseochromogenes | 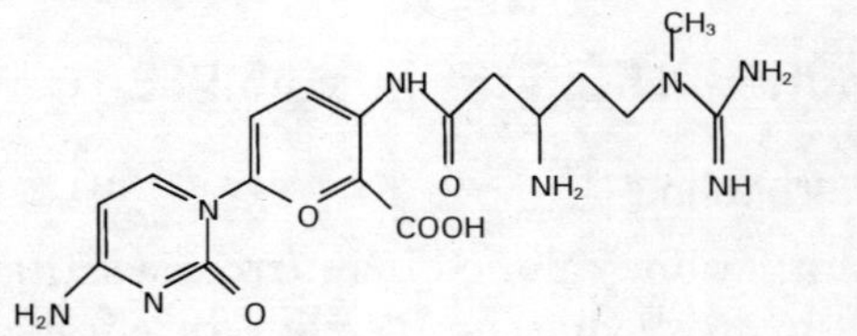 **USE:** Rice blast and inhibits certain bacteria |

## MISCELLANEOUS FUNGICIDES

### Dodine

(N-dodecylguanidine acetate)

Dodine is synthesised by treating dodecyl amine, cyanamide and acetic acid

$$C_{12}H_{25}\,NH_2 + NH_2CN + CH_3\,COOH \longrightarrow \left[C_{12}H_{25}\,NH\overset{NH}{\overset{|}{C}}—NH_3\right]^+ CH_3COO^-$$

Dodine has been used as a protective fungicide. It is recommended for the control of a number of major fungal diseases of apple, pear, vegetables and ornamental crops. Its acute oral $LD_{50}$ value for rats is 1000 mg/kg.

**Glyodine** is used as protective and curative fungicide against apple scab. The $LD_{50}$ for glyodine for rats is 3720 mg/kg.

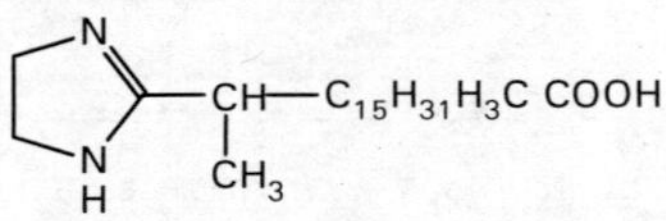

## BIBLIOGRAPHY

1. Thomson W.T., Agricultural Chemicals, IV, Fungicides Indianpolis, Indiana, 1973.
2. Horsfall J.G., Principles of Fungicidal Action, Chronica Botanica Company, Waltham, Mass, 1956.
3. Martin H., The Scientific Principles of Crop Protection Arnold, London, 1964.
4. Büchel K.H. ed Chemistry of Pesticides, Wiley, 1983.
5. Gruzdyer G.S., Mir Publishers, Moscow, 1983.
6. Matolcsy Gy, Nadasy M., Andrisha V., Pesticide Chemistry, Elseric, New York, 1988.

# CHAPTER 6

# Nematicides

Nematicides are chemical which prevent, repel, inhibit or destroy nematodes. Nematodes are multicellular animals and are found in nearly every biological niche that supports life. The nematodes are too small to be seen without the aid of microscope. Nematodes are also called eelworms, nemas, roundworms, threadworms, todes or worms.

## IMPORTANCE

Nematodes do not kill plants but they cause a reduction in the growth and vitality of the plant leading to an economic loss. Moreover they also serve as vectors for viruses, bacteria and fungi.

Nematodes are spreads primarily by man. The most common methods include the movement of infected plants or plant parts and infested soil adhering to implements. In India the plant parasitic nematodes are reported to cause considerable damage to variety of crops like tobacco, potato, vegetable crops, plantation crops, fruit crops, pulses, oilseeds, cereals etc.

The major plant parasitic nematodes in India are:

1. *Heterodera* sp
2. *Meloidogyne* sp
3. *Praytenchus* sp
4. *Radopholus* sp
5. *Rotylenchus* sp
6. *Helicotylenchus* sp

## PARASITISM

Parasitism of plants by nematodes is of two broad types:

### Ectoparasitism and Endoparasitism

Nematodes that live externally and pierce plant tissues with their stylet or spear or enter with the anterior portion of their body for the purpose of feeding are ectoparasites. Those, which embed their entire body within plant tissues and feed within are endoparasites.

Some genera infect and damage primarily the above ground plant parts and are some times inferred to as aerial parasites. These species may carry out their life cycle in the soil or in the shallow layers, when suitable host plants develop and favourable conditions prevail they attack the growing seedlings. The wheat gall nematode, the stem nematode and chrysanthemum foliar nematode are well known examples.

## Methods of Nematode Control

1. **Heat:** Excellent control of plant parasitic nematodes can be achieved through the use of heat in various forms. Temperatures necessary for effective control will depend on:
   a) other biological agents to be controlled
   b) soil conditions whether the soil is wet or dry
   c) exposure time.
2. **Hot water treatment:** Plants infected with nematodes are also controlled by hot water.
3. **Sanitation and use of Nematode free planting stock:** Sanitation and use of nematode free planting stock are sure and effective means of nematode control
4. **Time of planting and harvesting:** Certain parasitic nematodes are inactive during winter months, because low temperature inhibit their activities.
5. **Removal or destruction of infected host plants:** The root systems of certain annual crops will continue to live for several weeks after harvest and consequently parasitic nematodes present in the roots continue to reproduce. Populations of some nematodes can be reduced if the stalks are cut and the root systems are turned out soon after harvest.
6. **Fallowing:** Keeping the land free of all vegetation for varying periods by frequently tilling of the soil either by discing, ploughing or harrowing or application of herbicides to prevent plant growth results in starvation and death of nematodes present in soil.
7. **Nutrition and general care of host:** Nematode damage to certain crops can be minimized to some degree, by providing adequate nutrition and moisture and protecting the crop from adverse conditions such as drought or cold which place the plant under stress.
8. **Biological control:** Certain fungi capture and kill nematodes in the soil.
9. **Antagonistic crops:** In recent years considerable interest has been focussed upon plants with roots containing nematicidal substances. These toxic substances reduce the population level of some namatode species. African marigolds (*Tagets sp.*) and *Asparagus officinalis* L. are examples of such plants.
10. **Use of chemicals:** The chemicals used for control of nematodes, called nematicides are now recognised as an effective method of nematode control. Several effective nematicides have been developed and improvements in method of application have been made. In many parts of the world today soil treatment has become an established farm practice.

### Advantages of Chemical Control

1. Chemical control reduces population of nematodes to a very low density within a matter of days after chemical is applied.
2. Several nematicides used today are multipurpose. In addition to nematodes, fungi, bacteria and weeds are also controlled.

### Essentials of a Good Nematicide

1. Nematicides must have high vapour pressure to spread through the soil and to contact nematodes in the water films surrounding soil particles.
2. Nematodes are covered with an impermeable cuticle which provides them with considerable protection. Chemicals used as nematicides should have ability to penetrate the lipophillic cuticle.
3. Nematicides should have high water solubility.

### Nematicides are Classified into Five Groups

1. Halogenated hydrocarbons
2. Organophosphates
3. Isothiocyanates
4. Carbamates or oxime nematicides
5. Botanicals

## HALOGENATED HYDROCARBONS

### Methyl Bromide

Methyl bromide is prepared as follows:

1. By the reaction of methanol with bromine and hydrogen sulphide

$$2CH_3OH + Br_2 + H_2S \longrightarrow 2CH_3Br + 2H_2O + S$$

2. By reaction of methanol and salts of HBr in the presence of $H_2SO_4$

$$CH_3OH + NaBr + H_2SO_4 \longrightarrow CH_3Br + H_2O + NaHSO_4$$

Methyl bromide is phytotoxic and it must be used as a preplant application, followed by adequate aeration time. It is effective in controlling stem and root knot nematodes.

### Ethylene Dibromide

Ethylene dibromide is produced by direct bromination of ethylene

$$CH_2{=}CH_2 + Br_2 \longrightarrow BrCH_2CH_2Br$$

EDB is injected into the soil several days before planting to kill nematodes, eggs and soil insects.

### 1,3 Dichloropropene

It is obtainted as by product by chlorination of allyl chloride which is obtained by chlorination of propene

$$CH_3CH{=}CH_2 + Cl_2 \longrightarrow ClCH_2CH{=}CH_2 + HCl$$

$$Cl\,CH_2\,CH{=}CH_2 + Cl_2 \longrightarrow ClCH_2CH{=}CHCl + HCl$$

### 1,2-dichloropropane

1,2- dichloropropane is prepared by chlorination of propylene at low temperature

$$CH_3{-}CH{-}CH_2 + Cl_2 \longrightarrow CH_3{-}\underset{Cl}{CH}{-}\underset{Cl}{CH_2}$$

### DD

(Nemafene)

The technical product is a clear amber liquid containing at least 50% of cis and trans isomers of 1,3-dichloropropene ($CHCl{=}CH{-}CH_2Cl$) which is the main active ingredient and 50% of 1,2- dichloropropane ($CH_3{-}CHCl{-}CH_2Cl$) and other trichlorides.

DD is extremely toxic nematicide and injection into soil kills both eelworms and their cysts.

### Nemagon

(1,2 dibromo-3 chloropropane)

Nemagon is prepared by the addition of bromine to allyl chloride

$$CH_2 = CH\,CH_2\,Cl \xrightarrow{Br_2} Br\,CH_2\,CH\,Br\,CH_2\,Cl$$

DBCP is used as post plant treatments and is registered for use on fruits and vegetables and is less toxic than DD. In 1977 it was announced that DBCP was found to cause sterility in male workers involved in its manufacture and subsequently it has been removed from the market. It is currently available only as restricted use pesticide for a very few essential needs.

## ISOTHIOCYANATES

### Metham Sodium

(Vapam, sodium N-methyl dithiocarbamate)

Metham -sodium (Vapam) is prepared by the reaction of methylamine with carbon disulphide in presence of sodium hydroxide

$$CH_3\,NH_2 + CS_2 + NaOH \longrightarrow CH_3\,NH\,\overset{\overset{S}{\|}}{C}\,SNa + H_2O$$

Metham-sodium is a preplant fumigant effective against nematodes, fungi and weeds. The toxic action of metham-sodium is explained by the evolution of methyl isothiocyanate that disrupts the oxidation reduction processes in the cells of organisms.

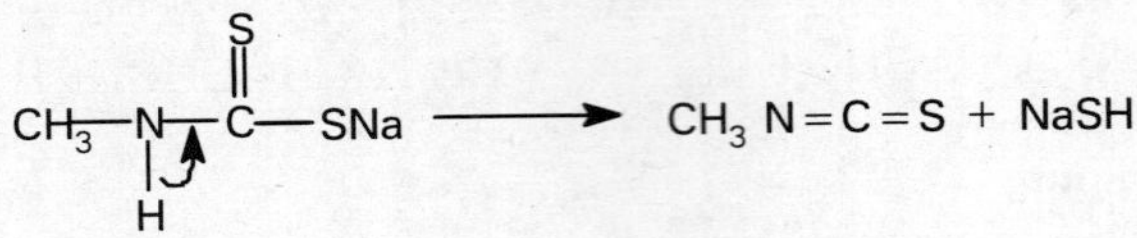

## Dazomet

(Tetrahydro-3,5-dimethyl-2H-1,3,5-thiadiazine-2-thione)

Dazomet is prepared by the reaction of carbon disulphide, methylamine and formaldehyde

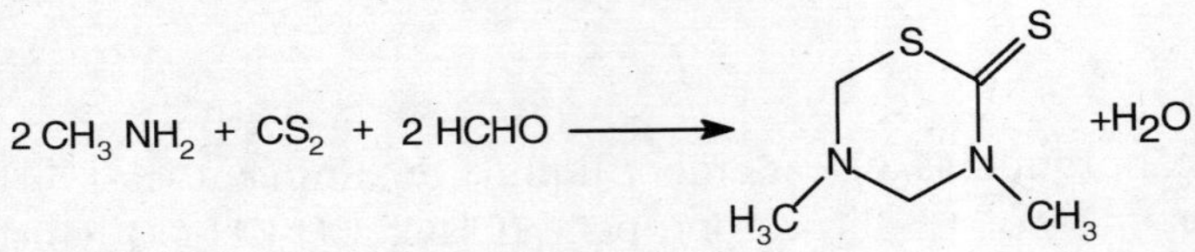

Dazomet like metham-sodium owes its activity due to its decomposition in the soil to give methyl isothiocyanate

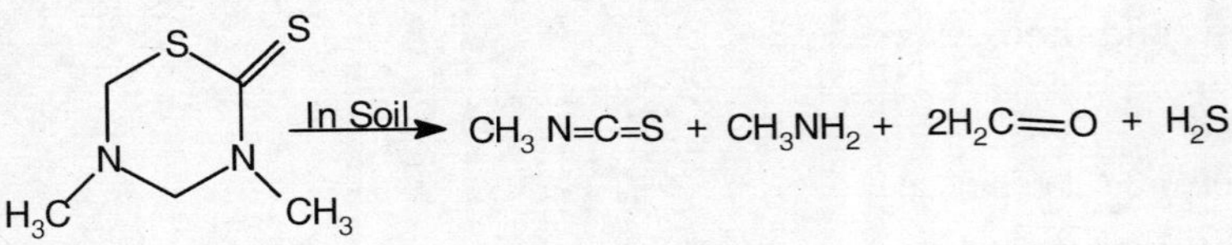

## Organophosphates

Nematodes have nervous system similar to insects. All these organophosphates inhibit the nervous transmitter enzyme, cholinesterase resulting in paralysis and ultimately death of the affected nematode. Following organophosphate nematicides are used for killing nematodes.

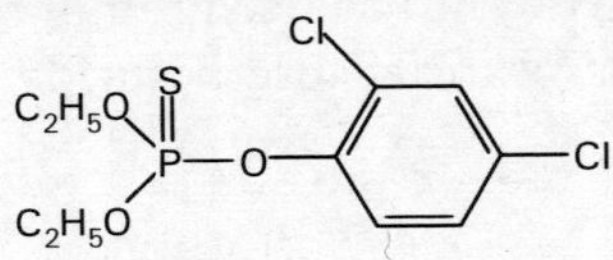

Dichlofenthion

(O,O diethyl dichlorophenyl phosphorothioate)

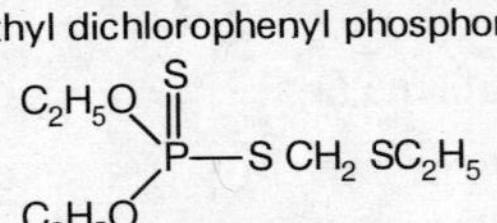

**Phorate**

(O,O diethyl -S- (ethylthiomethyl) phosphorodithioate)

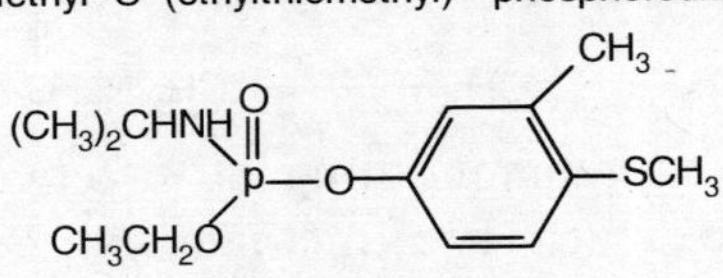

**Fenamiphos**

(Ethyl -3-methyl -4-(methylthio) phenyl - (1-methyl ethyl) phosphoramidate)

## Carbamates

Several carbamates are also highly nematicidal and examples include carbofuran and aldicarb. These are highly active against soil nematodes but suffer from very high mammalian toxicity.

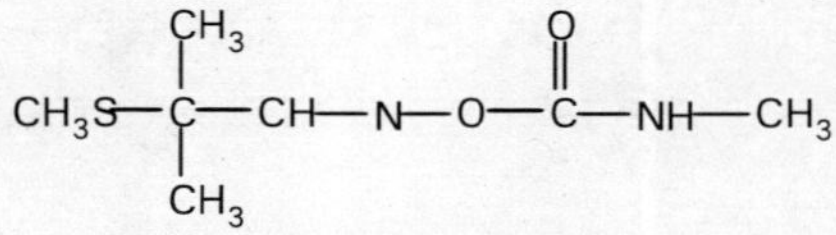

Adicarb (Temik)
(2 - methyl - 2 - (methylthio) propionaldehyde O- (methylcarbamoyl) oxime)

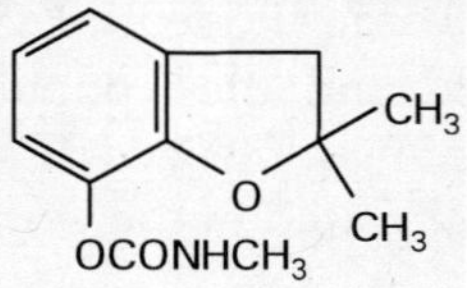

Carbofuran (Furadan)
(2,3 - dihydro - 2,2 - dimethyl - 7 - benzofuranyl methylcarbamate)

## Botanical Nematicides

Non edible oil cakes like Neem, Mahua and Karanj control nematode population. Marigold (*Tagetis erecta* L) has been found effective in controlling nematodes. The nematicidal principles of root exudates of marigold have been identified as terthienyl (I) and substituted bisthienyl compound (II)

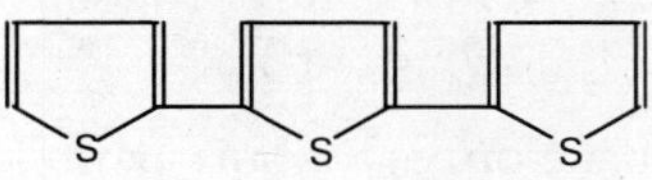

(I)

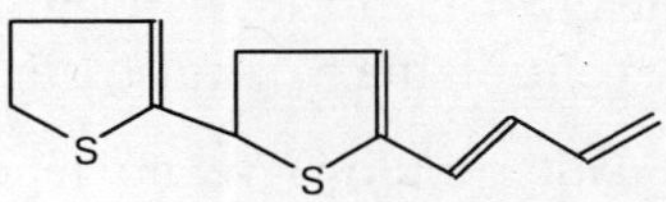

(II)

## BIBLIOGRAPHY

1. Metcalf R.L. Advances in Pest Control Research. Interscience New York 1960.
2. Gruzdyer G.S., The Chemical Protection of Plants, Mir Publishers, Moscow 1983.
3. Matolesy Gy., Nadasy, and Andriska K Pesticide chemistry, Elsever, 1988.

# CHAPTER 7

# Rodenticides

Rodenticides are chemicals that prevent, inhibit, destroy, kill or mitigate rodents (class-mammalia, order-Rodentia). More than 750 species of rats live throughout the world and their population, estimated to exceed 3 billions, constitutes a permanent menace through diseases and destruction.

The main species of rats found in the world are as follows:

- ***Rattus norvegicus***: *Rattus norvegicus* accounts for more than 95% of all the existing rats and is known world wide as the largest species. It lives close to mankind and expels the other species.
- ***Rattus rattus***: *Rattus rattus* is mainly found in warmer region and in industrial buildings such as mills and warehouses.
- ***Mus musculus***: *Mus musculus*, the house mouse is less important pest economically but lives in close association with man in his dwellings
- ***Rattus alexandrinus***: *Rattus alexandrinus* is mainly found in ponds and near the water front.

Generally rodenticides are applied as baits to control rodent pests. For effective control, the bait must satisfy the following criteria.

1. The bait should have no repellent action or odour. Bait should be atractive, clean, fresh and hard.
2. It must not induce bait shyness.
3. The poison should make rats go out into the open to die, otherwise the rotting corpses create health hazards.
4. It should have much lower toxicity to domestic animals especially cats and dogs which eat the poisoned rats.

Redenticides can be classified into following groups:

## INORGANIC COMPOUNDS

White arsenic, yellow phosphorous, zinc phosphite, arsenic trioxide have been used as rodenticides.

### Botanicals

Red squill which comes from the powdered bulbs of a plant *Urginea maritima* is used as rodenticide. The active ingredient is an alkaloid scilliroside (1).

Strychnine (2) another alkaloid from *Strychnos nuxvomica* is generally converted into strychnine sulphate and is used as rodenticide

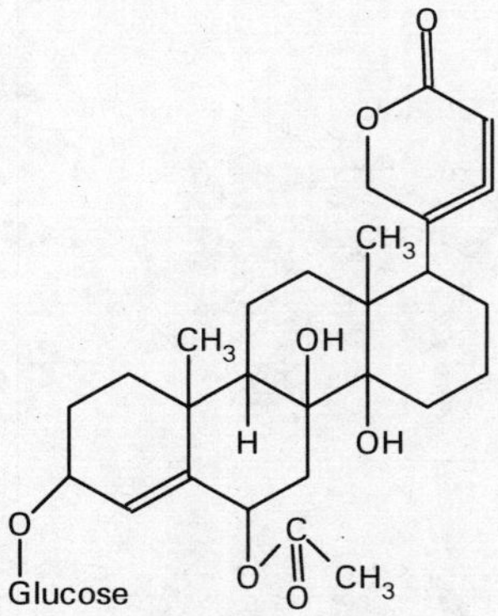


(i)
Scilliroside

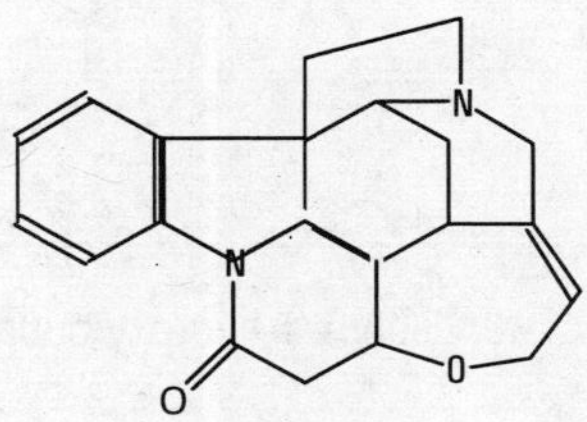


(ii)
Strychnine

## COUMARINS

Coumarins are the most important group of rodenticides. Most of the coumarins act as anticoagulants.

### Warfarin

(3-(1- phenyl-2-acetylethyl)-4-hydroxy coumarin).

WarFarin is prepared by condensation of 4-hydroxycoumarin and benzalacetone.

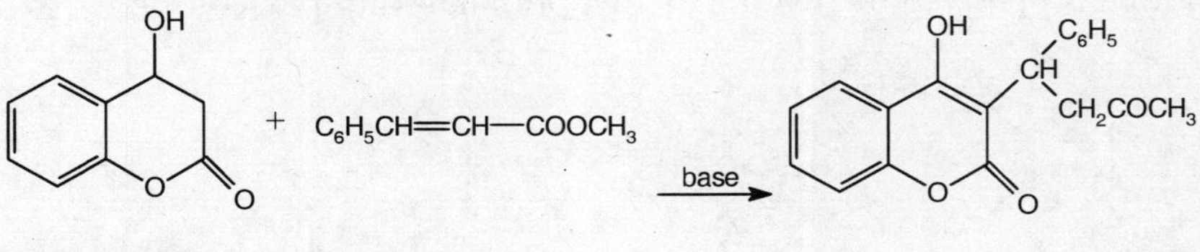


### Coumatetralyl

(3-(1,2,3,4-tetrahydro napthyl) 4-hydroxy coumarin)

Coumatetralyl is prepared by condensation of 4-hydroxy coumarin with - hydroxytetraline in acetic acid solution

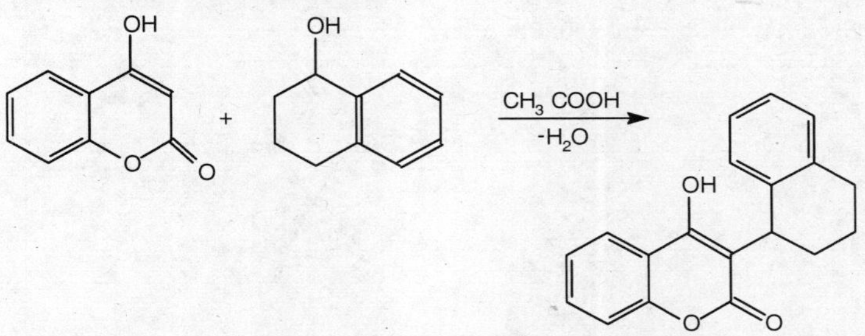


The other rodenticides which contain coumarin group and act as

rodenticides are:

**Coumachlor**

3-(2-acetyl-1-p-chlorophenylethyl)-4- hydroxy coumarin

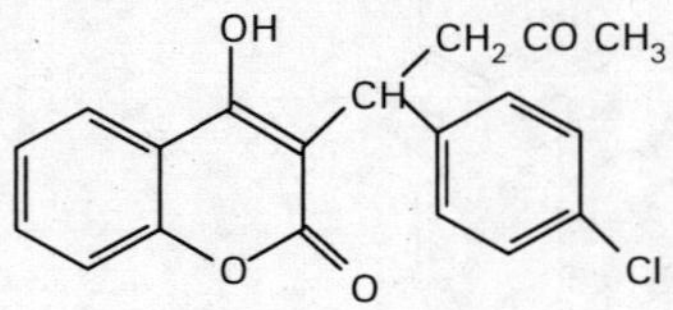


**Coumafuryl**

(3-(2-acetyl 1-furylethyl)-4-hydroxy coumarin)

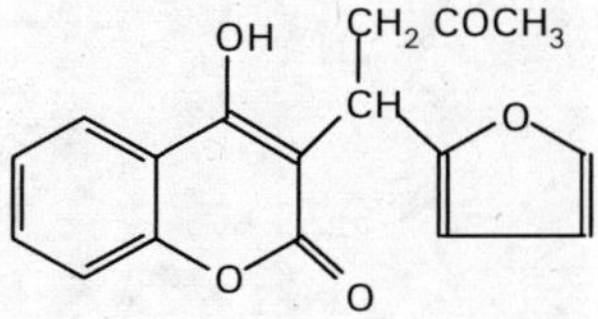


## INDANEDIONES

These substituted indanediones have also remarkable anticoagulant property. Examples of this class of rodenticides are

**Pindone**

(2-pivaloylindane-1,3-dione)

Pindone is prepared by treating diethyl phthalate and pinacolone

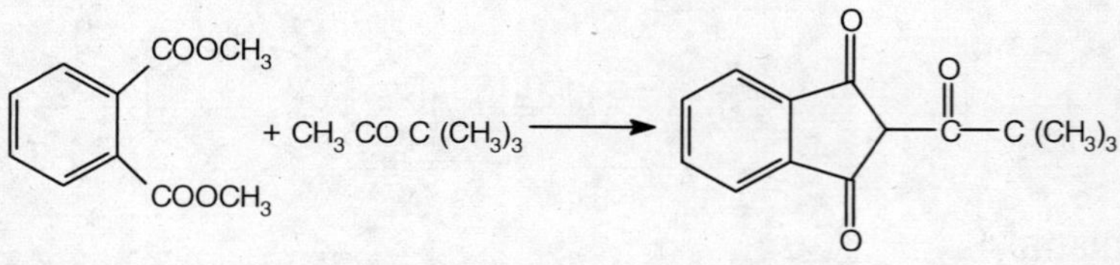


Other compounds which contain Indanedione nucleus and act as rodenticide are:

**Diphacinone**

((2-diphenyl acetyl indane-1,3-dione)

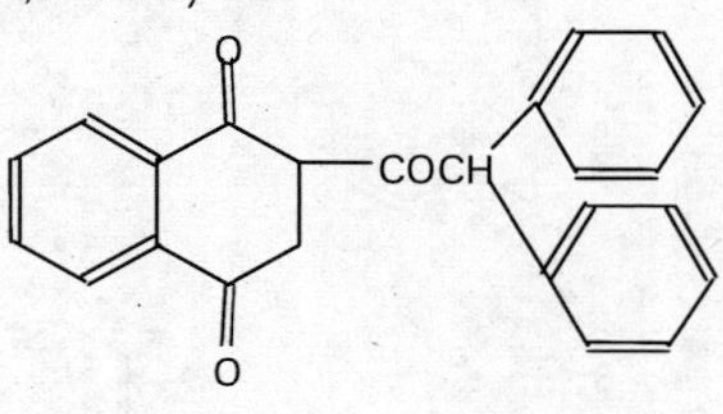

## Chlorophacinone

((2-p- chlorophenyl-2-phenylacetyl) indane-1,3-dione)

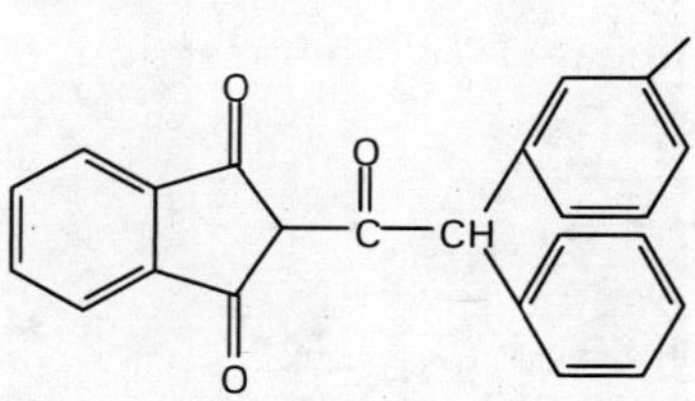

# ORGANOPHOSPHOROUS

## Gophacide

(O,O-bis (p-chlorophenyl) acetimidoylphosphoramidothioate

It is prepared by condensation of O,O-di-p-chlorophenyl phosphorochloridothioate with acetamidine

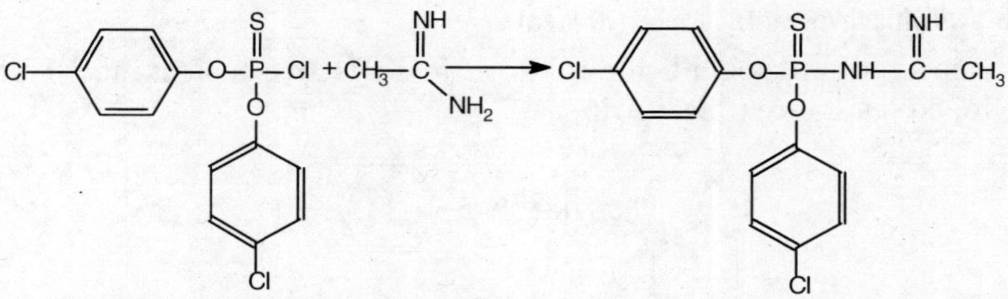

Gophacide is very effective for control of a range of rodent pests, but prove hazardous to non-target species.

# CHLORINATED RODENTICIDES

## Crimidine

(2-chloro-N,N,6-trimethyl-4-pyrimidinamine)

Crimidine is prepared by condensation of urea and ethylacetoacetate followed by chlorination and dimethylation.

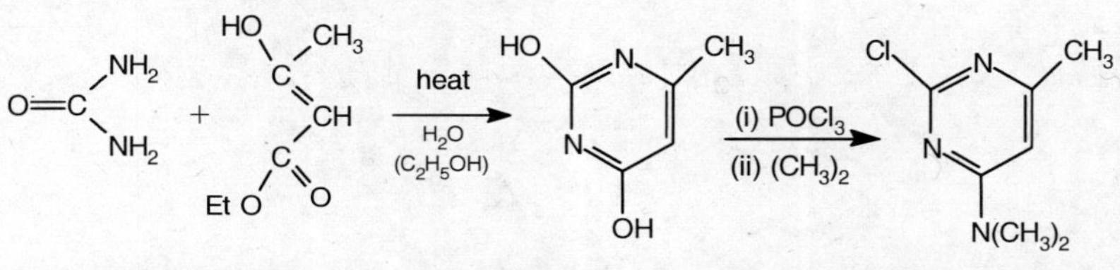

Crimidine is highly toxic to rats and is readily metabolized in the body, therefore poisoned rats are non-toxic to predators.

## MISCELLANEOUS RODENTICIDES

### α-Napthylthiourea or ANTU

α-Napthylthiourea is prepared by the reaction of α-napthylamine and ammonium thiocyanate

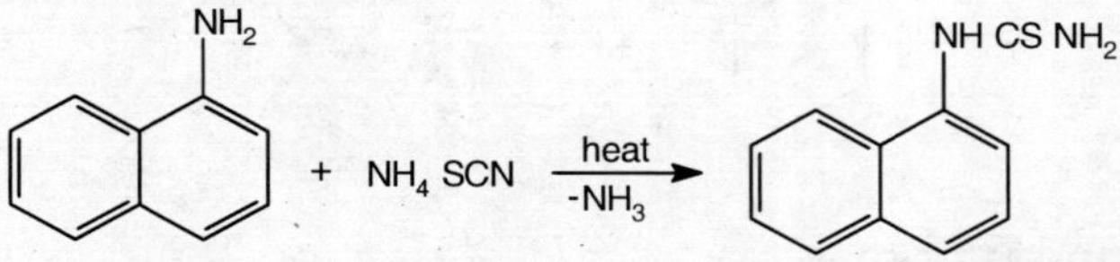


ANTU is a specific poison for adult Norway rats. ANTU is banned in some countries because of carcinogenicity of occassional impurities such as α-napthylamine. As rodents develop a tolerance to ANTU., it has been replaced by other materials.

### Vacor

(1-(3-pyridylmethyl)-3-(4-nitrophenyl) urea)

Another important rodenticide is Vacor. It is effective against rats that have developed resistance to warfArin.

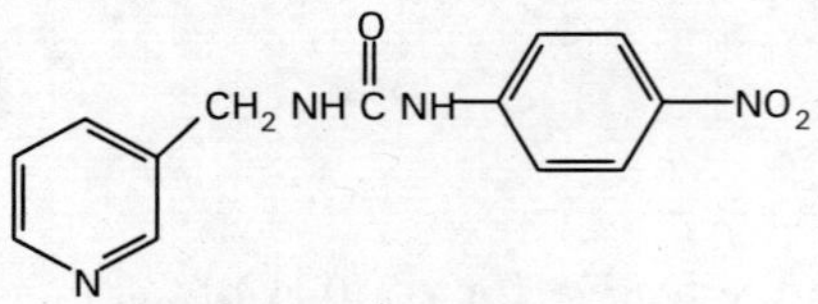


### Brodifacoum

(3-(3-(4'-bromobiphenyl-4-yl)-1,2,3,4-tetrahydronapth-1-yl)-4- hydroxy coumarin)

Brodifacoum is effective against resistant rodents and its effectiveness against Norway rats and roof rats is more than that of any other rodenticide.

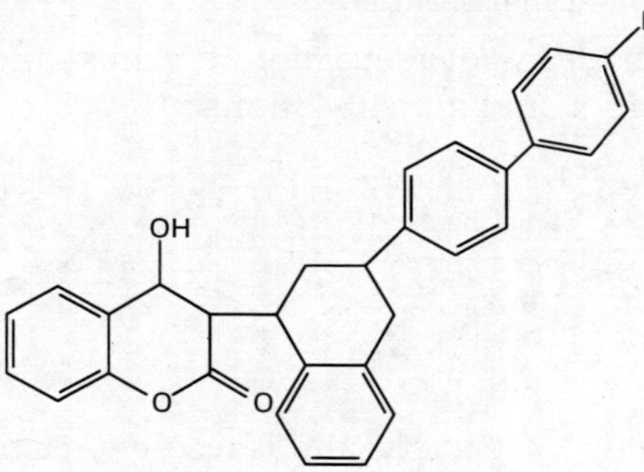


### Bromadiolone

(3(3-(4'-bromo-1,1'-biphenyl-4-yl)-3- hydroxy-1-phenyl propyl)-4-hydroxycoumarin)

Bromadiolone is highly palatable anticoagulant and capable of controlling Norway rats

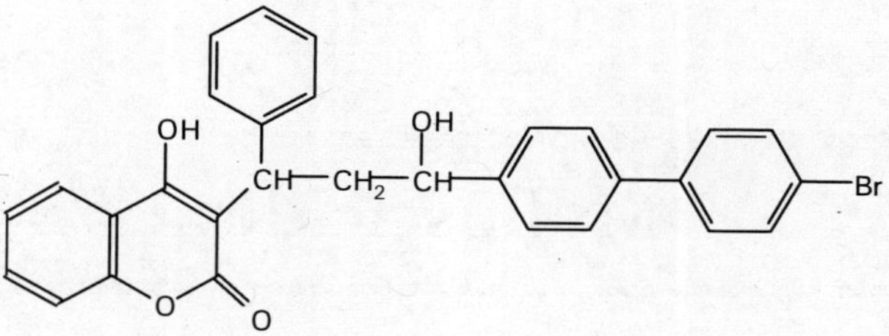


**Norbormide**

(Raticate, (5-(α-hydroxy- α-2-Pyridylbenzyl)-(7-(α-2- pyridylbenzylidene)-5-norbornene-2,3-dicarboximide)

It is highly selective rodenticide and is non toxic to other animals.

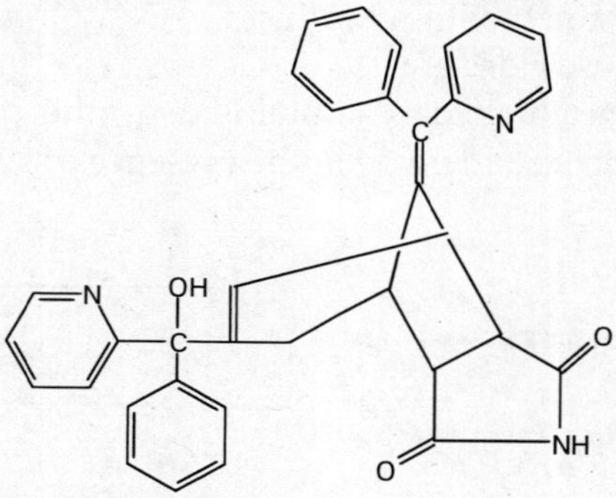


## PROBLEMS IN RODENT CONTROL

Both in the field as well as in the building, a number of rodent species live in association with each other in India as against single species composition in Europe and in other countries. The toxicities of rodenticides differ from species to species with the result that the predominant species is soon controlled but the others survive unaffected, so the choice and evaluation of rodenticides should be therefore of wide spectrum nature, highly cost effective, safe to non-target species, easy to handle and which is effective under diferent ecological situations. In India the main hinderances to rodent control are the religious and social taboos of the peace loving farmers. Therefore there is need for creating awareness of rodent control in a large number of farm families to combat rodent menance. Huge losses are caused by rodents every year to cereals, millets, pulses, sugarcane, cotton, vegetables, fruits, cocca, coconut, eggs and to almost all agricultural products.

## BIBLIOGRAPHY

1. Bücheled K.H. chemistry of pesticides Wiley 1983.
2. Shepard H.H., The chemistry and action of insecticides Mac Graw - Hill, New York, 1951.
3. Peardon D.L., Pest control 42, 14, 1974.

# CHAPTER 8

# Molluscicides

Molluscs include land slugs and snails which do a considerable amount of damage by eating vegetable seedlings and green vegetables. Molluscs do not directly harm mammals, but are alternate hosts for some species of parasites. Mulluscicides are chemicals which prevent, repel, inhibit or destroy members of the Phyllum Mollusca.

Important molluscicides are:

**Metaldehyde** Metaldehyde is obtainted by the polymerization of an ethanolic solution of acetaldehyde in the presence of sulphuric acid

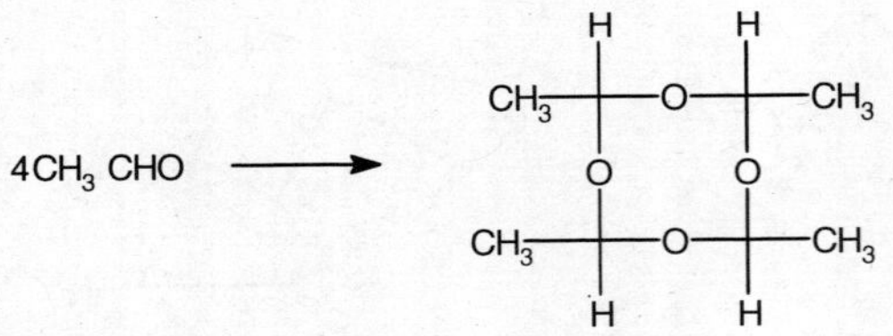

Metaldehyde has been used as baits for the control of slugs and snails. Its continued use can be attributed to its attractant and toxicant qualities

## Methiocarb

(3,5-dimethyl-4-(methylthio) phenyl methyl carbamate)

It is prepared by the reaction of 4-methyl thio-3, 5-xylenol and methyl isocyanate

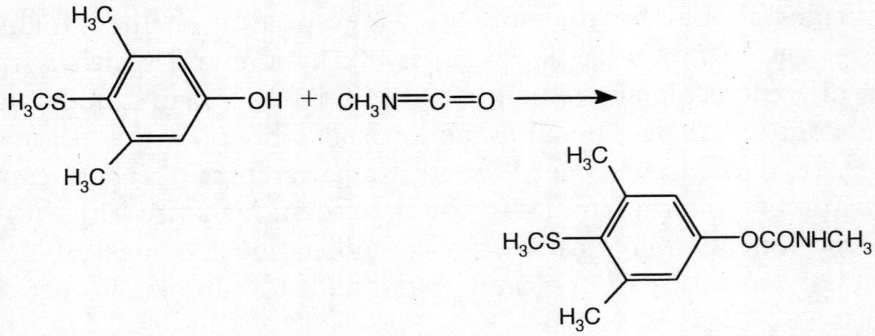

Methiocarb is registered for use against snails and slugs on ornamental plants.

## Trifenmorph

4-(triphenyl methyl) morpholine

Trifenmorph is obtainted by the condensation of triphenyl methyl chloride with morpholine

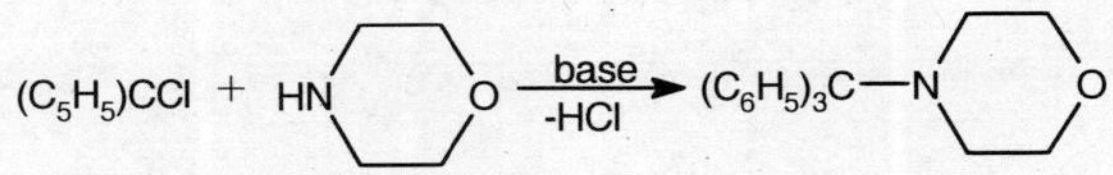


Trifenmorph is used against aquatic snails.

## Bayluscid

(5-chloro-N-(2-chloro-4-nitrophenyl)-2-hydroxybenzamide)

Bayluscid is prepared by condensation of 5-chlorosalicylic acid and 2-chloro-4-nitroaniline

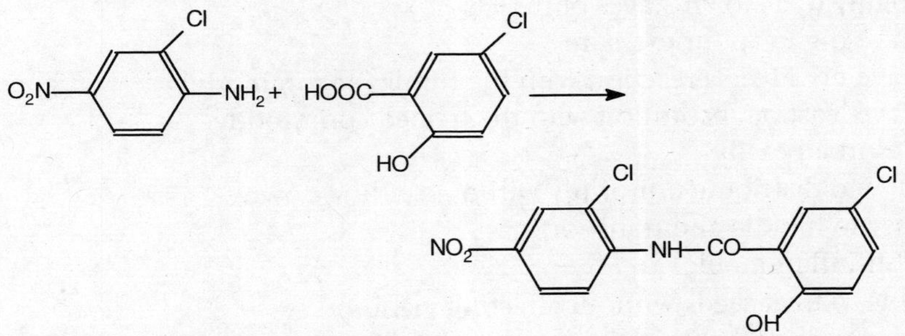


Bayluscid is formulated as the water soluble ethanol amine salt and is used against snails.

## TRIPHENYLTINACETATE

Triphenyltinacetate is prepared by treating triphenyltinhydroxide with acetic acid

$$(C_6H_5)_3\ SnOH + CH_3COOH \longrightarrow (C_6H_5)_3\ SnOCOCH_3$$

Triphenyltinacetate is used as a fungicide, algicide and moulluscicide.

**Copper sulphate** ($CuSO_4$) has also been used to kill the snail, liver fluke vectors.

## BIBLIOGRAPHY


1. Buchel H.K. ed. chemistry of pesticides, Wiely, 1983.
2. Gruzdyer G.S., Mir Publishers, Moscow, 1983.
3. Thomson W.T. Agricultural Chemicals; Thomson Publications, India, 1975.

# CHAPTER 9

# Fumigants

A fumigant is a chemical which at the required temperature and pressure exists in a gaseous state in sufficient concentration to be lethal to a given pest or organism.

Important characteristics of an ideal fumigant:

It should be:

1. Highly toxic to all stages of insects
2. Have high vapour pressure
3. Have no adverse reaction with the fumigant commodity.
4. Have least of absorption with the treated commodity.
5. Desorbed easily
6. Able to distribute uniformly without settling down
7. Easy to handle and transport
8. Non-inflammable
9. Able to treat seeds without affecting viability
10. Be economical and indigenously manufactured.

## FACTORS INFLUENCING FUMIGATION

Through respiration insects directly develop susceptibility to fumigants since a lethal dose is absorbed by their tissues in a shorter or longer period. The rate of respiratory metabolism is increased by increase in temperature, increase in carbon dioxide content of fumigated space, decreases in oxygen content of fumigated space and also increase in the action on pests. Some other factors like density or insect population, pre and post fumigation effects, starvation effects, relative humidity, sub-lethal fumigation length or exposure period, fumigation interval and relative susceptibility of diferent stages of insects also play important role in fumigation.

Important fumigants are:

Ethylene dichloride carbon tetrachloride mixture (3:1 by volume)

$$Cl\,CH_2\ CH_2\ Cl\,CCl_4$$

The EDCT mixture is a popular fumigant due to least hazards involved in its use, inspite of many drawbacks. Being heavy, gas diffusion is slow and requires prolonged ventilation after fumigation. The carbon tetrachloride is feared to induce cancer and therefore not favoured in some countries.

## ETHYLENE DIBROMIDE

It has high boiling point 131.6°C and is sorbed by many fumigated commodities resulting in poor penetration. It is a nerve poison and a narcotic. It causes serious injury to the central nervous system and needs careful handling.

## METHYL BROMIDE

This is a very effective fumigant with a gas tight structure for controlling most insects. Because of its low boiling point and high volatile nature, it is readily converted into gas, rendering it for fumigation. In pure form it is a colourless liquid boiling at 3.6°C with a density of 1.732 g/cm$^3$. It is non-explosive and non-inflammable. It is a difficult fumigant to handle.

Methyl bromide decomposes in mammals to form methyl alcohol and subsequently formaldehyde which incresses the toxic action of the fumigant.

$$CH_3Br + H_2O \longrightarrow CH_3OH + HBr$$

$$CH_3OH + H_2 \longrightarrow HCHO + H_2O$$

### Methallyl Chloride

(1- chloro-2- methylprop-2-ene)

Methallyl chloride is a fumigating insecticide and acaricide. It is highly toxic to a broad range of stored grain pests. The fumigant in the form of vapour readily penetrates into a pile of grain to depth of two meters. It is a colourless oily liquid boiling point 72°C and is highly toxic to human and warm blooded animals.

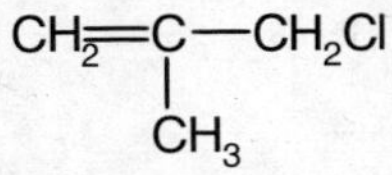

Methallyl chloride does not corrode metals and does not destroy rubber and other materials.

### Chloropicrin

(trichloronitromethane)

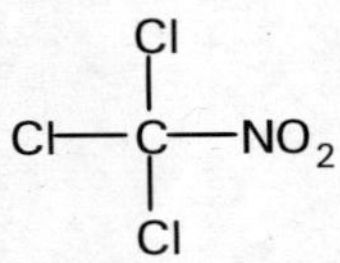

In pure form it is colourless heavy liquid boiling at 112.4°C. Its volatility at 20°C is 16 mg/litre. Chloropicrin is sorbed to a great extent by grains and is slowly desorbed.

### Aluminium Phosphide

It is a solid fumigant in compressed tablet or pellet forms. The round tablet is dark brown in colour and yellowish tinge. It is composed of pure aluminum phosphide and ammonium carbamate. The tablet consists of a minimum 56% w/w aluminum phosphide (active ingredient) and maximum 44% w/w inert ingredients. The tablet weighs three gram and releases about one gramme of phosphine on exposure to moisture. In addition, carbondioxide and ammonia are also evolved, thus preventing self inflammation of phosphine gas. It is indigenously manufactured with indigenous raw materials and the cost of fumigation is less.

### Ethylene Oxide [$(CH_2)_2O$]

It is a gas and is marketed as a liquid in pressure cylinders. It is mainly employed in grain stores and for plant disinfection.

### Formaldehyde (HCHO)

Formaldehyde is used as 40% aqueous solution and is mostly used as surface dressing for seed beds to combat soil fungi.

Other fumigants include methyl isocyanate, hydrocyanic acid and sodium N-methyl dithiocarbamate dihydrate.

$$CH_3\,NCS \;;\; HCN \;;\; CH_3\,NH\,\overset{\displaystyle S}{\overset{\|}{C}}-\overset{\displaystyle S}{\overset{\|}{C}}\,SNa\;\; 2H_2O$$

## BIBLIOGRAPHY


1. Hassal K.A., World Crop Protection Pesticides, London Iliffe Books, 1969.
2. Brown A.W.A., Insect control by chemicals John Wiley and sons, New York, 1951.
3. Peardon D.L., Pest control 42, 14, 1974.
4. Gratz N.G. Bull W.H.O. 48 469, 1973.

# CHAPTER 10

# Acaricides

Acaricides are chemicals used for control of mites that damage plants.

## CHLORINATED COMPOUNDS

### Dicofol
### (1,1-bis (chlorophenyl)-2,2,2-trichloroethanol)

Dicofol is prepared by chlorination of (chlorfenethol)

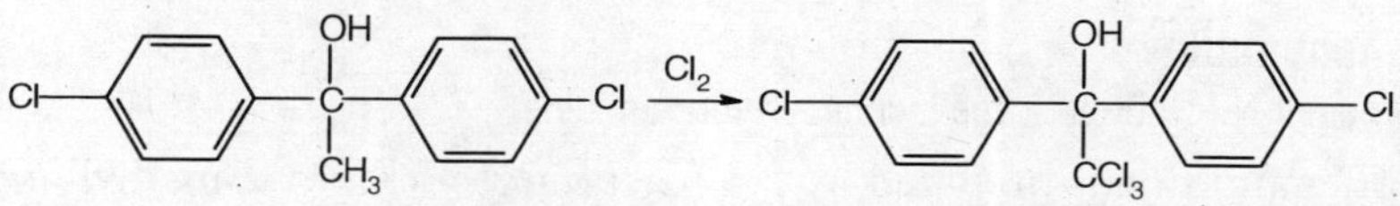


Dicofol is a contact acaricide with a high initial toxicity to phytophagus mites. Its acute $LD_{50}$ for rats is 430 mg/kg.

### Dienochlor
(bis (pentachloro-2,4- cyclopentadien-1-yl)

Dienochlor is obtainted by catalytic hydrogenation of hexachloro-cyclopentadiene

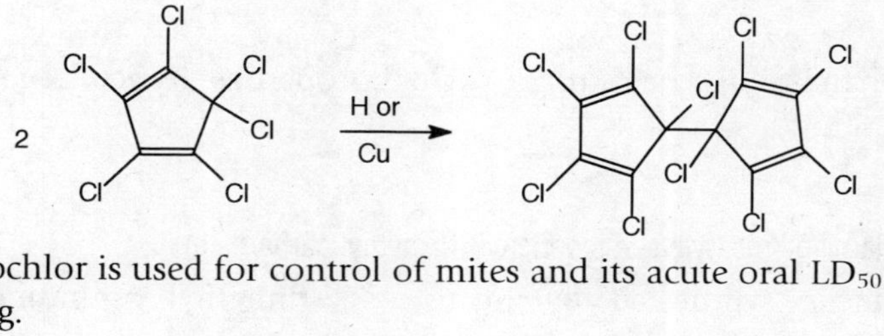


Dienochlor is used for control of mites and its acute oral $LD_{50}$ for rats is 316 mg/kg.

### Chlorfenethol
(1,1-bis (4-chlorophenyl) ethenol)

Chlorfenethol is prepared by treating 4,4'-dichlorobenzophenone and methyl magnesium bromide.

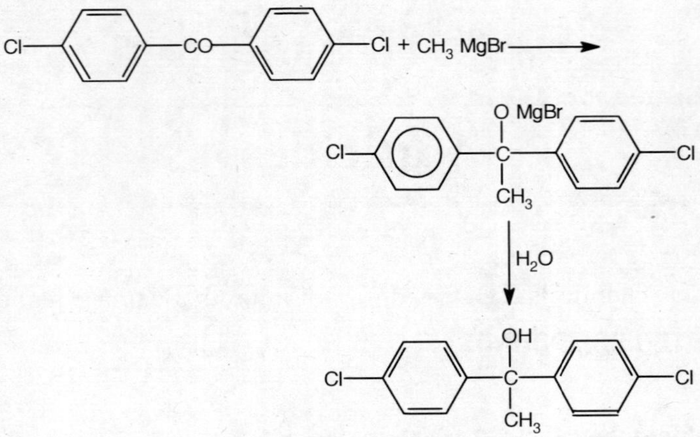

## DERIVATIVES CONTAINING A C=N OR N=N DOUBLE BOND

### Chlorphenamidine

(N,N-dimethyl-N (4-chloro 2-methylphenyl) formamidine)

Chlorphenamidine is prepared by the reaction of 4-chloro-2-methyl-aniline and dimethyl formamide.

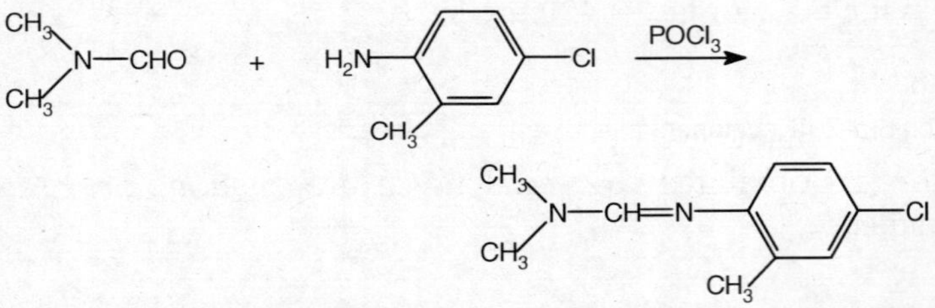

Chlorphenamidine has strong ovicidal properties. Its acute oral toxicity to rats is 335 mg/kg.

### Formetanate

(3,3-dimethylamino-(methyleneiminophenyl) N-methyl carbamate)

Formetanate is synthesied by treating 3,3-dimethyl aminomethylene-iminophenol with methyl isocyanate

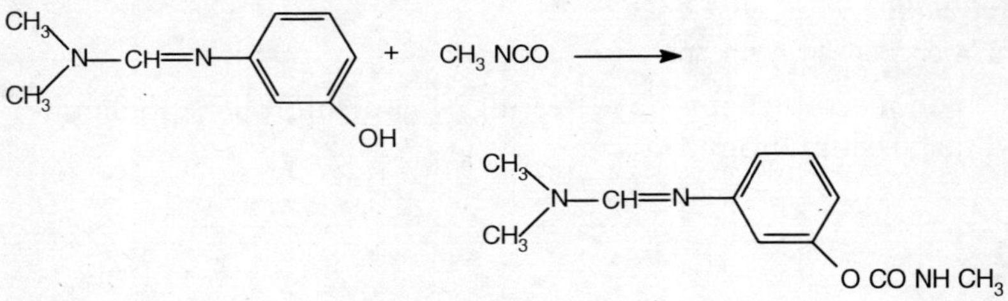

Formetanate is effective against spider mites. The acute oral $LD_{50}$ of formetanate for rats is 24 mg/kg.

## COMPOUNDS CONTAINING SULPHUR

### Chlorfensulfide

(4-chlorophenyl-2,4,5-trichlorophenylazosulphide)

Chlorfensulfide is prepared by reaction of 2,4,5-trichlorobenzene-diazonium chloride with 4-chlorothiophenol

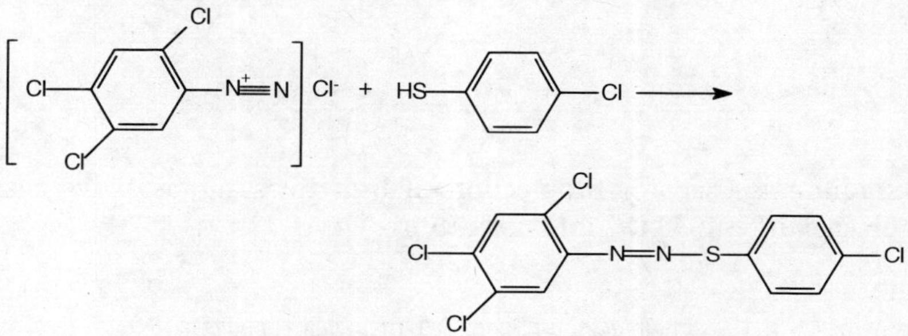

Its acute oral $LD_{50}$ for mice is 3000 mg/kg.

### Tetrasul

(4-chlorobenzyl-2,4,5-tri-chlorophenyl sulphide)

Tetrasul is prepared by treating 4-chlorothiophenol with 1,2,4,5-tetra-chlorobenzene

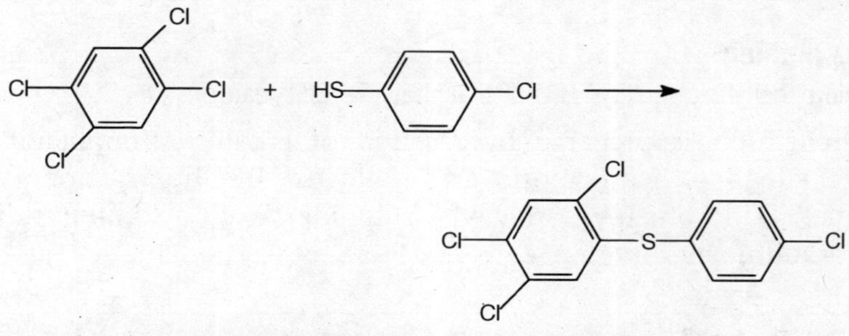

Tetrasul is toxic to tetranychis mites. Its acute oral $LD_{50}$ for female rats is 6800 mg/kg.

### Tetradifon

(1,2,4-trichloro-5(4-chlorophenyl sulfonyl) benzene)

Tetradifon is prepared by friedel-crafts reaction between 2,4,5-tri-chloro benzenesulfonyl chloride and chlorobenzene.

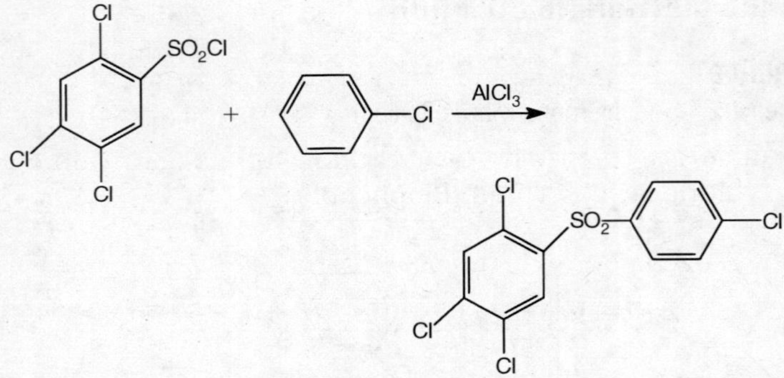

Tetradifon is used for proptection of fruit trees and is active against phytophagus mites. Its $LD_{50}$ for rats is 15 mg/kg.

### Aramite

(2-chloroethyl 2-(p-1,1-dimethylethyl) phenoxy) 1-methylethyl sulfite)

It is used as acaricide.

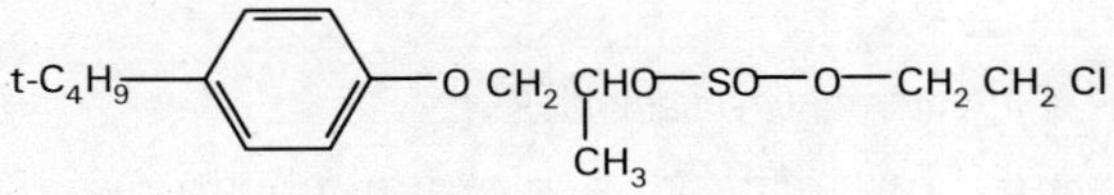

## ORGANOPHOSPHOROUS COMPOUNDS

### Carbophenethion

(S-((4-chlorophenylthio) methyl) 0,0-diethyl phos phorodithioate)

Carbophenethion is prepared by reaction of p-chloro thiophenol with formaldehyde and hydrochloric acid followed by the reaction of the substituted chloromethyl phenyl sulphide with sodium diethyl phosphorodithioate.

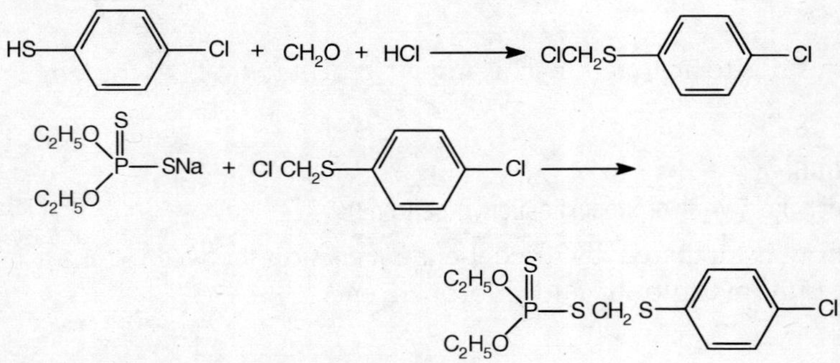

Its $LD_{50}$ value for rats is 28-100 mg/kg.

## ORGANOMETALLIC COMPOUNDS

### Fentin Hydroxide

(Triphenyltin hydroxide)

It is used against plant feeding mites. Its acute oral $LD_{50}$ for rats is 540 mg/kg.

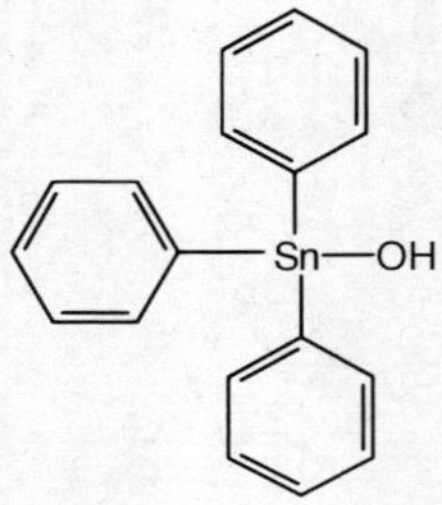

## NITROPHENOL DERIVATIVES

### Binapacryl

(2-1-methylpropyl)-4,6-dinitrophenyl-2-methyl-2-crotonate)

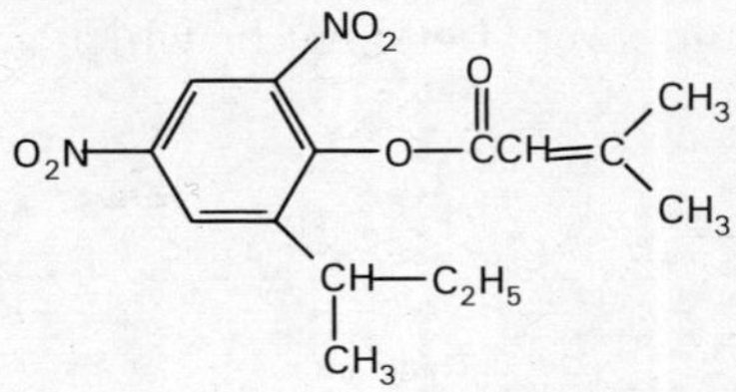

### Dinobuton

(1-methylethyl 2-(l-methyl propyl) 4,6-dinitrophenyl carbonate)

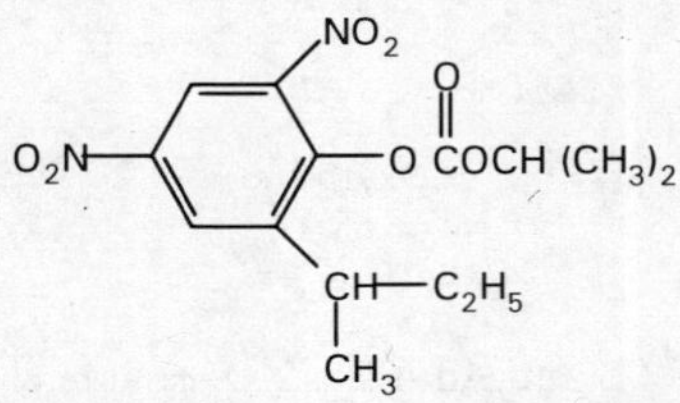

Binapacryl and Dinobuton are mainly effective against mobile stages of spider mites. The $LD_{50}$ of binapacryl and dinobuton to rats are 165 and 140 mg/kg respectively.

## FORMAMIDINES

### Chlordimeform

(N'-(4-chloro-2-methylphenyl) N,N-dimethylmethanimidamide)

Its acute oral $LD_{50}$ value to rats is 340 mg/kg rats.

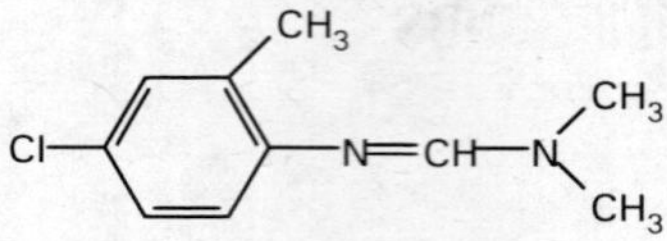


### Formetanate Hydrochloride

(N,N-dimethyl-N'- 3 (methylamino carbonyl) oxy) phenyl methanimidamide) monohydrochloride

Its acute oral $LD_{50}$ to rats is 21 mg/kg.

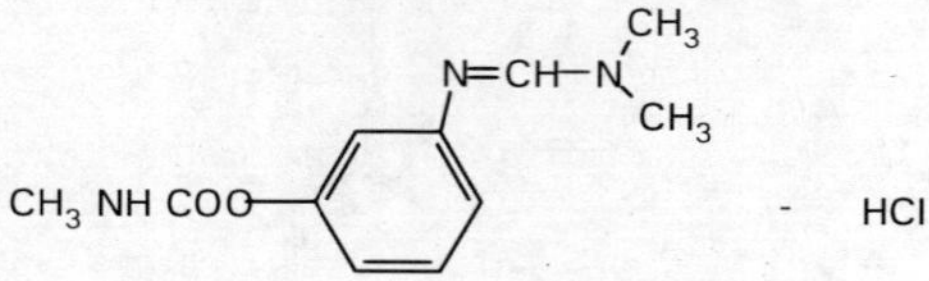


## HETEROCYCLICS

### Oxythioquinox

(6-methyl-2,3-quinoxalinedithiol cyclic dithiocar-bonate) and thioquinox (2,3-quinoxalinedithiol cyclic trithiocarbo-nate)

They are important examples of heterocyclic which act as acaricides.

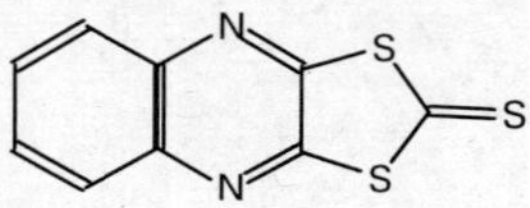


Thioquinox

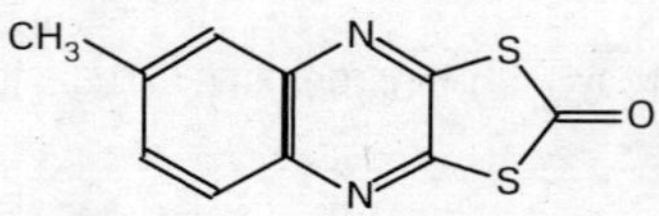


Oxythioquinox

## BIBLIOGRAPHY

1. Maltolscy Gy., Nadasy M. and Andriska V. Pesticide chemistry, Elserier, New York, 1988.
2. Brown A.W.A., Insect control by chemicals John Wiely and sons, New York, 1951.
3. Büchel K.H., Chemistry of Pesticides, Wiely, 1983.

# CHAPTER 11

# Herbicides

A herbicide in the broadest sense of the word, is any compound that is capable of either killing or severely injuring plants and may thus be used for elimination of plant growth or killing of plant parts.

The designation weed covers a variety of meanings according to a particular situation. According to the general definition weed is any plant either a wild or cultivated variety that is undesired in that particular place.

## How Weeds Cause Damage or Losses

1. Weeds decrease crop yields by removing a vast amount of water and nutrients from the soil.
2. They also compete with crops for space and sunlight.
3. Weed material in agricultural seed, in grain and hay reduce the value and quality of these products.
4. Many weeds cause loss to farmers by acting as host to insect pests and to other organisms which cause plant diseases.
5. Livestock may be poisoned by certain poisonous weeds that grow in the field or get into harvested crops. Poison milkweed, poison hemlock, pigweed are just a few examples.
6. Weeds may cause many other problems for man and animals. Plants such as sandbur, three-owned grass and wild barley have spines and harbed awns which may cause physical injury to the feet, mouths and eyes of animals.
7. The presence of weeds in crop increases wear on machinery.
8. The weeds compete for carbon dioxide required for photosynthesis.

## How Weeds are Introduced and Spread to New Areas

1. Weeds may be distributed by nature and man.
2. Nature, animals and birds help spread of weeds. The seeds of many plants have hooks, barbs or become sticky wet so they are made to cling to the wool or hair of animals. Both animals and birds eat many weed seed which can pass through their digestive tracts without being damaged. Seeds can thus be spread through the manure or dropings over large distances.
3. A strong wind can spread almost all kind of weed seeds.
4. Rains or melting snows are natural ways by which weeds may be spread

to long distances.

5. Runners, rhizomes, and creeper type of weed roots are capable of spreading weeds over large areas regardless of their seed production and disperse.
6. Impure farm seed is the important way by which man has aided in the introduction and spread of weeds.
7. Modern irrigation systems are able to collect and distribute water over large areas of land. This system also aids weed seeds to spread throughout this large area.
8. Farm machinery spreads weed seeds specialy in wet weather, when the seeds attach to implements and vehicles in mud or by their natural stickness.

## WEED CLASSIFICATION

Based on life span, weed species are classified into three broad groups annuals, biennials and perennials.

1. **Annuals:** Plants with a one year life cycle are annuals. They propagate through seeds. They produce a crop of seed and die. There are two types of annuals a) Summer annual b) Winter annual.
   a) Summer annuals are plants that result from seeds which germinate in the spring, grow, mature, produce seed and die before winter of each year. Examples are crabgrass, foxtail, pig weed.
   b) Winter annuals are plants that grow from seeds which germinate in the winter, produce flowers and ripens seeds in the following spring or early summer and then die. The seeds of winter annuals remain dormant in the soil during summer. Examples are annual blue grass and winter galic.
2. **Biennials:** Plants with a two year life cycle are biennials. These plants require two years to complete their life cycle. They grow from seeds which germinates in the spring. They develop a heavyroot and a compact rosette or cluster of leaves in the first summer. Biennials remain dormant through the first winter and in the second summer they mature, produce seed and die before winter. Examples *Cyanadon dactylon*.
3. **Perennials:** Plants which live more than two years and may live indefinitely are perennials. These plants may grow from seed, but many produce tubers, bulbs, rhizomes and stolons. The above ground parts of these plants may die back each winter but develop new above ground parts each spring. Examples Johnson grass, field bindweed, donoletion, and plantain. Perennial plants can be further grouped as follows:
   a) Simple perennials reproduce mostly by seeds but in some instances root pieces may produce new plants following mechanical injury during cultivation. Bulbous perennials propagate through the

underground parts like bulbs, and tubers as well as by seeds e.g. wild onions and wild garlic. Creeping perennials spread by lateral extensions of the creepers above ground stems (rhizomes) either by roots or by seeds. Examples are Bermuda grass, Johnson grass.

## METHODS OF WEED CONTROL

### Prevention

Prevention of the introduction of the weeds to new areas deals mainly with attempts to cut down the steps of spread of weeds caused by man.

1. Stopping the use of impure farm seed. There are laws regulating the quality of farm seed used.
2. Keep hay and grain as pure as possible so that animals do not further spread weeds from the seeds still alive in manure and droppings.
3. Keeping the farm mechinery and other equipments used in farming free of weed seeds.
4. Limiting the introduction of weed seeds through irrigation systems.

### Cultural and Mechanical Control

1. **Crop rotation:** A carefully planned rotation of crops involving hay, grain and intertilled crops is a practical and effecient method. Many weeds that are known to associate with certain crops are destroyed in this manner.
2. **Cultivation:** This is the oldest and most effective mechanical method of controlling weeds. However all weeds are mostly easily and economically killed when they are seedlings.
3. **Mowing:** This is a most effective method of preventing the maturation of annual and biennial seeds. Mowing is used often in places where cultivation is impossible. Flowering should not be permitted before mowing because many weeds are able to mature seeds after the flowers have been grown even if they are cut.
4. **Grazing:** This method serves the same purpose as does the method of mowing.

### Biological Control

In biological weed control a 'natural enemy' of the pest plant is used which is harmless to all other plants.

Parasitic plants or highly competitive plants are used as biological control. The risk of introducing new insect plant or disease into a new area is great. Much time must be spent on testing of the proposed control organisms to see first of all if it can survive in a new environment and secondly it destroy only the noxious weed in field.

### Chemical Weed Control

Employing chemicals for weed control constitutes chemical method. These chemicals, commonly referred to as weedicides or herbicides, are being used for over 85 years.

Benefits of chemical weed control:

1. Herbicides can be applied for weed control in crop rows where cultivation is impossible.
2. Preemergence herbicides provide early season weed control. This is very beneficial as crop-weed competition during early stages of crop growth is maximum than at later stages.
3. Cultivation and manual method of weed control may injure the root system in plantation crops like tea and coffee which grow for several decades.
4. Herbicides can control many perennial weed and brush species which cannot be controlled by other methods.

## CLASSIFICATION OF HERBICIDES

The herbicides are classified on the basis of

1. method of application
2. mode of action
3. chemical affinity and structural similarity

### Classification Based on Method of Application

Classification on the basis of method of application divides herbicides into two main groups. Soil applied or preemergence and foliage applied or post emergent herbicides.

Preemergence are applied to soil before crops and weeds emergence. It may also refer to applications after crops emergence but before the weeds emerge. Presence of adequate soil moisture is essential for the success of pre-emergence herbicides.

Post emergence are applied after the emergence of crop and weeds.

Preplant herbicides are generally applied before planting of crops.

### Classification Based on Mode of Action

The classification on the basis of mode of action of herbicides takes into account differences in the physiological and biochemical action of herbicides. On this basis they are categorized as systemic or translocated herbicides and non systemic or contact herbicides.

1. **Contact herbicides:** Herbicides that kill plant parts when come in contact are known as contact herbicides. They must be directly toxic to living cells and the effects are usually rapid. Contact herbicides are effective against annual weeds, but usually only 'burn off'perennial

plants. Selective contact herbicides kill or stunt some plants with little or no injury to others.

2. **Translocated Herbicides:** Herbicides that can be absorbed by either the roots or above ground parts and move or are translocated through the plant sytems. These herbicides upset the plant's growth and metabolic processes particularly enzyme systems. Full effects may not be evident until a month or more after application.
3. **Residual:** A chemical which provides weed control for a period of time even after the application as soil applied herbicide. Residual chemicals may be selective or non-selective. A selective herbicide is a herbicide that is more toxic to some kinds of plants than to other. The degree of selectivity is affected by plant age, rate of growth, form of plant and physiological differences. Non selective herbicide is toxic to all plants.

## Classification Based on Chemical Affinity and Structural Similarity

The herbicides which originate from one chemical nucleus have structural and chemical similarities and affinities and thus provides the basis for chemical classification. The two major classifications are inorganic and organic herbicides.

**Inorganic Herbicides:** Inorganic herbicides include copper sulphate ($CuSO_4.5H_2O$), copper nitrate ($CuNO_3$), ammonium sulphate ($(NH_4)_2SO_4$), ammonium sulfonate ($NH_4$-$SO_3$-$NH_2$), sodium tetraborate ($Na_2B_4O_7.10H_2O$), sodium metaborate ($Na_2B_2O_4.4H_2O$), amorphous sodium borate ($Na_2B_4O_{13}.4H_2O$), sodium chlorate ($NaClO_3$), ammonium thiocyanate ($NH_4SCN$) and sulphuric acid ($H_2SO_4$). These were used between 1896 and 1930's. Intensive EPA restrictions have been placed on some inorganic herbicides because of their persistence in soil. The inorganic herbicides are not wise choice for use except aiming to remove all vegetation from an area.

## The Organic Herbicides

*Petroleum oils:* The earliest organic herbicides were the petroleum oils which are a complex mixture of long chain hydrocarbons containing traces or nitrogen and sulphur linked compounds. These mixtures contain alkanes and often alicyclics and aromatics. The petroleum oils are effective herbicides for all vegetation. The petroleum oils exert their lethal effect by penetrating and disrupting plasma membranes.

*Organic arsenicals:* Organoarsenicals are widely used as agricultural herbicides. They are the organic arsenicals namely arsenic and arsenic acid derivatives. The following arsenic herbicides are used in agriculture.

| | | |
|---|---|---|
| 1. | Methanearsenic acid is used for control of grasses in cotton and turf. | 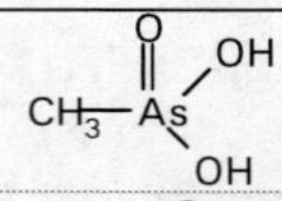 |
| 2. | Disodium methane arsonate is used for control of grass weeds in cotton and rubber plantations | 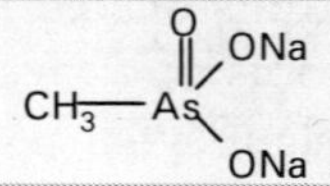 |
| 3. | Monosodium methanearsonate is used in tea and rubber plantation. | 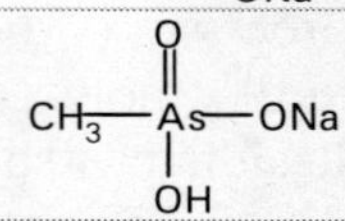 |
| 4. | Cacodylic acid (dimethyl arsenic acid): Cacodylic acid is used as postemergence contact herbicide for controlling weeds in non cropped areas. It is also used as defoliant and desiccants for cotton. | 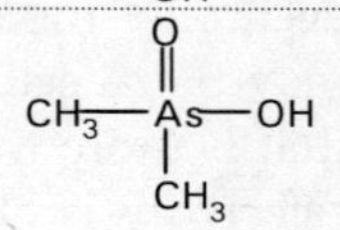 |
| 5. | Hexaflurate (Potassium hexafluoro-arsenate): Hexaflurate is a selective post-emergence herbicide for the control of prickly pear in non cropped areas | 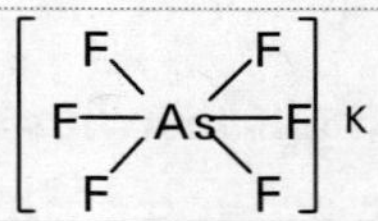 |

**Aliphatic**

| | | |
|---|---|---|
| 1. | Acrolein is used as aquatic herbicide on submerged and floating weeds mainly in irrigation canals, detacher drains and ponds. | 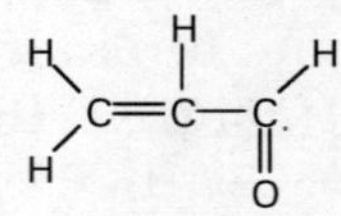 |
| 2. | Allyl alcohol used as a preplant treatment to kill weed seeds in tree nurseries, tobacco seed beds, and ornamental plantings. | 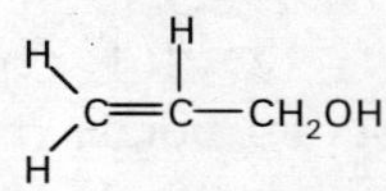 |
| 3. | Trichloroacetic acid controls wild oat and is effective against perennial grasses. | 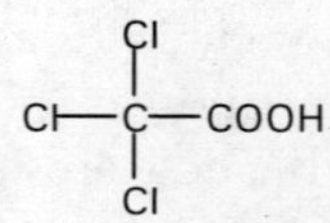 |
| 4. | Dalapon is applied as post emergence on foliage and used for control of weeds in sugarcane, grapes, coffee, tea and orchard crops. | 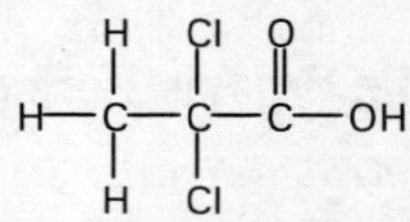 |

## PHENOXYALIPHATIC ACIDS

These herbicides are highly selective for broad leaf weeds and are translocated through the plant system.

### 2,4-D

(2,4-dichlorophenoxyacetic acid)

2,4-D is a selective systemic postemergence herbicide in used for the control of many annual broadleaf weeds in cereals, sugarcane and plantation crops. It is available in salt and ester formulations. Ester formulations are volatile. 2,4-D is prepared by the following methods:

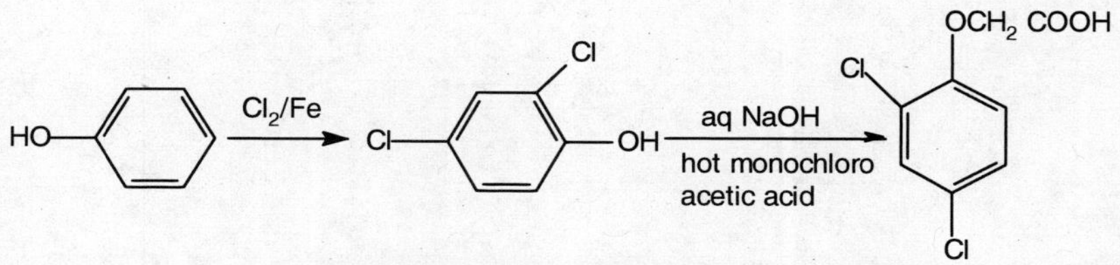

Chlorination of phenoxy acetic acid

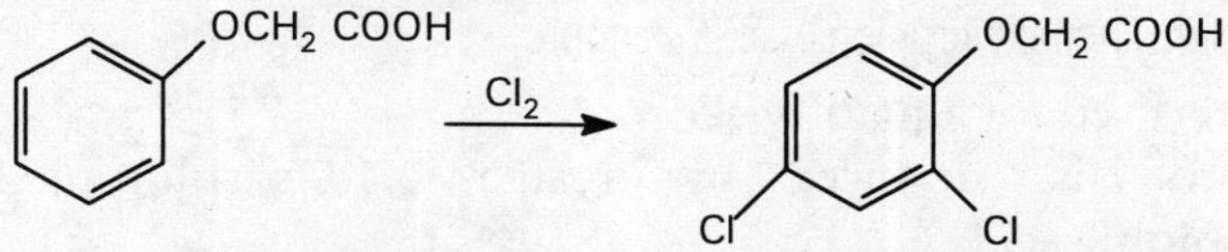

The esters are formed from 2,4-D with respective alcohols in the presence of catalyst. Isopropylester is most widely used.

## MCPA

(4-Chloro-2-methylphenoxyacetic acid)

MCPA is more selective than 2,4-D at equal rates. It is widely used in cereals legumes and aquatic areas. It is available as amines, esters and sodium or potassium salts.

It is prepared by treating 4-chloro-2-methyl phenol with monochloroacetic acid in alkaline medium by heating the reaction mixture to 103-105°C.

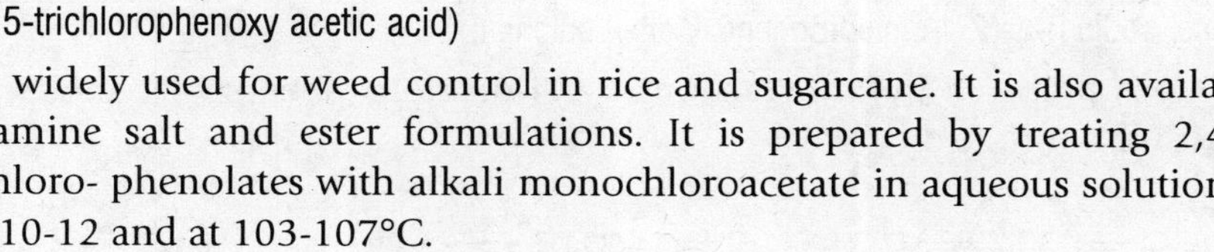

## 2,4,5-T

(2,4,5-trichlorophenoxy acetic acid)

It is widely used for weed control in rice and sugarcane. It is also available as amine salt and ester formulations. It is prepared by treating 2,4,5-trichloro- phenolates with alkali monochloroacetate in aqueous solution at pH 10-12 and at 103-107°C.

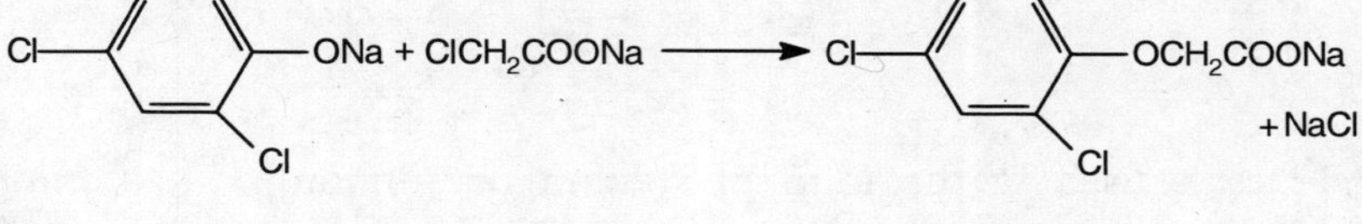

## MCPP

(4-chloro-2-methylphenoxy propionic acid, Mecoprop)

MCPP is prepared by the condensation of p-chloro-O-cresol with -chloro propionic acid.

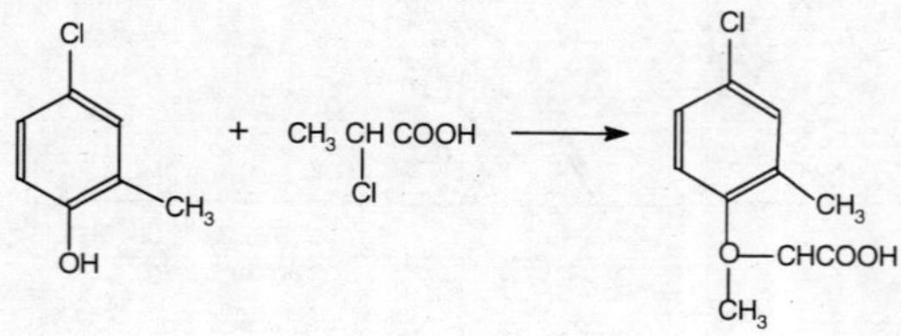

## 2,4,5-TP

(2,4,5- trichlorophenoxy propionic acid, Fenoprop)

Fenoprop is used for aquatic weeds.

Fenoprop is produced by treated 2,4,5-trichlorophenolate with chloropropionic acid.

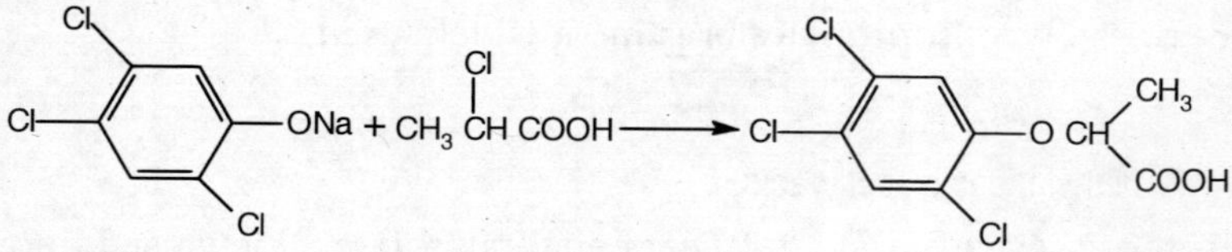

## MCPB

(2-methyl-4-chlorophenoxy butyric acid)

MCPB is prepared by reacting sodium 4-chloro-2-methylphenolate with 1,3-chlorobromopropane

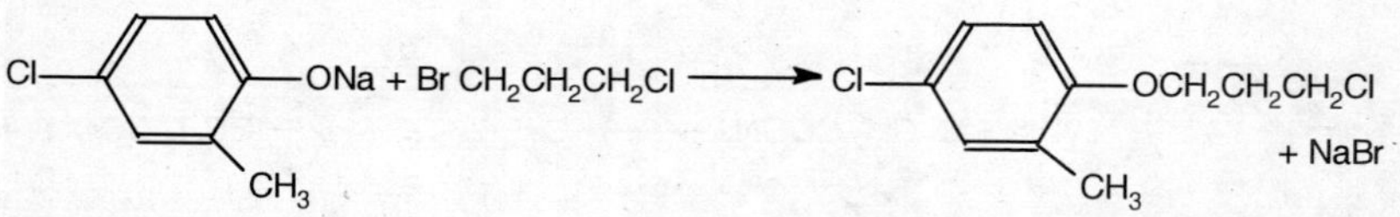

MCPB is used for control of weeds in legume crops.

## 2,4-Des

(sesone, sodium-2-(2,4-dichlorophenoxy ethyl sulphate)

2,4-Des is prepared by the reaction of 2,4-dichloro phenoxy-ethanol with chlorosulphonic acid

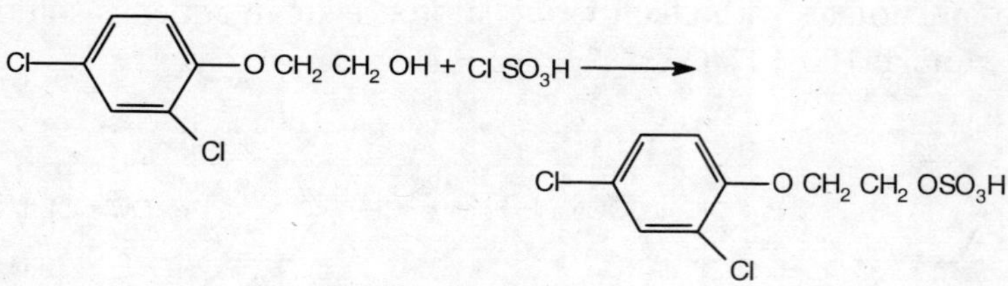

2,4-Des is used in the form of sodium or ammonium salt for the control of broad leaved and grassy weeds.

## Erbon

{2-(2,4,5-Trichlorophenoxy) ethyl 2,2-dichloropropanoate)

Erbon is prepared by the esterification of 2,4,5-trichlorophenoxyethanol with 2,2-dichloro propionic acid in the presence of sulphuric acid.

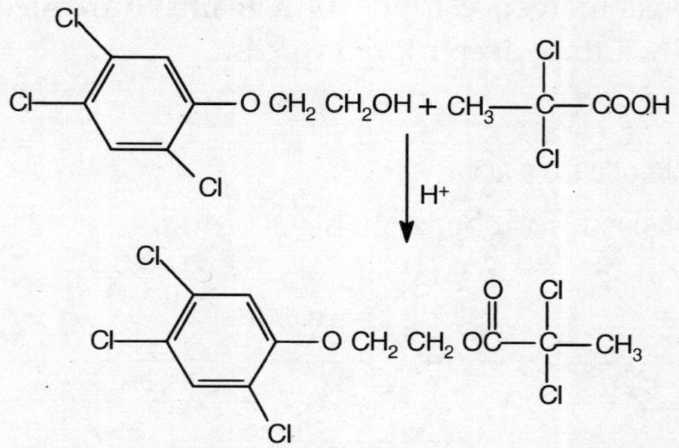

Erbon is a non-selective herbicide used for killing vegetation in noncropped areas.

## 2,4-DEP

(tris [2(2,4-dichlorophenoxy) ethyl] phosphite and
bis [2-(2,4-dichlorophenoxyethyl)] phosphite)

It is prepared by reacting phosphorous trichloride and 2,4-dichlorophenoxyethanol in the presence of pyridine.

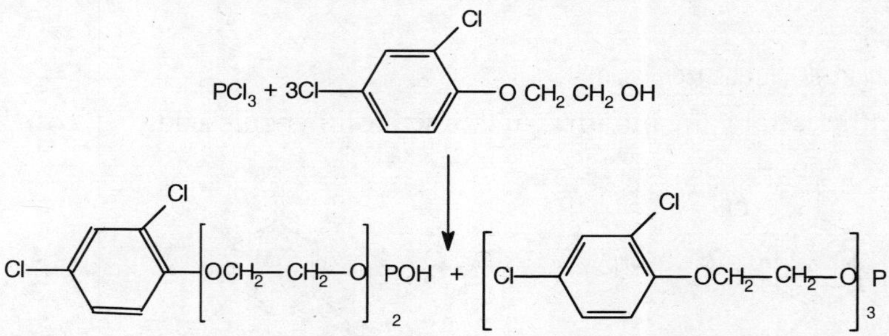

It is a selective herbicide effective for the control of mono-and dicotyledonous weeds in maize, potato and peanut. It is hydrolysed in soil to 2,4-dichlorophenoxyethanol.

These herbicides are mainly used for annual and perennial grass weeds in broad leaved crops.

# DERIVATIVES OF BENZOIC ACID

## 2,3,6-TBA

(2,3,6-trichlorobenzoic acid)

Trichlorobenzoic acid is obtained by chlorination of benzyl chloride in the presence of a catalyst

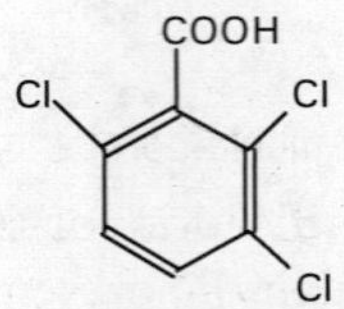

It is used as postemergence application against broad-leaved annual and perennial weeds including deep rooted weeds.

## Dicamba

(2-methoxy-3,6-dichlorobenzoic acid)

Dicamba is synthesised from benzene hexachloride.

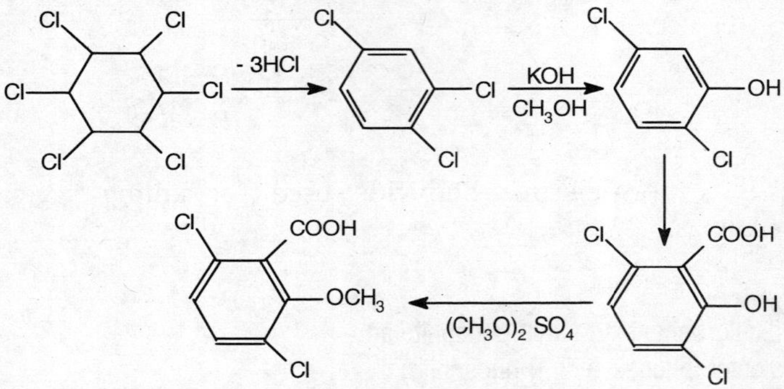

Dicamba, in the form of its dimethylamine salt is used for control of grasses, broad leaved weeds and several shrubs.

## Dinoben

(2,5-dichloro-3-nitrobenzoic acid)

It is synthesised by the nitration of 2,5-dichlorobenzoic acid.

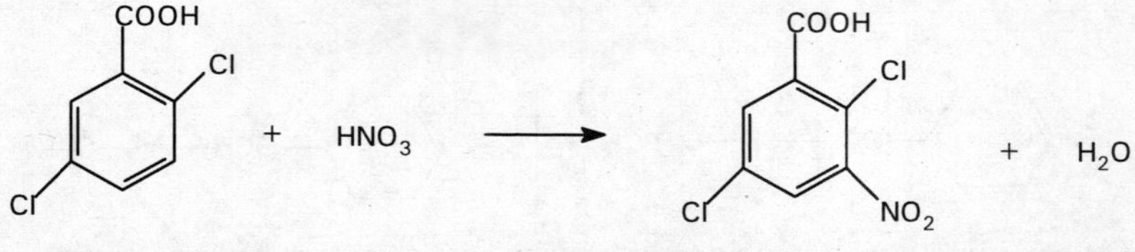

Its water solution, sodium and trimethyl amine salts are preemergence selective weed killers against annual and broad-leaved weeds in maize, soybean and horticultural crops.

# BENZONITRILES

## Dichlobenil

(2,6-dichlorobenzonitrile)

Dichlobenil is obtained by converting 2,6-dichlorotoluene in the presence of catalyst vanadium pentaoxide with mixture of ammonia and oxygen.

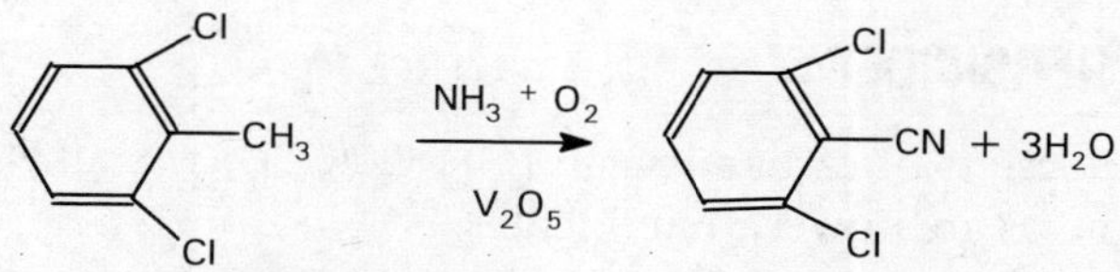

Dichlobenil is effective against several annual and perennial weeds. Dichlobenil can be used for the control of aquatic weeds.

### Chlorthiamid

(2,6-dichlorothiobenzamide)

Chlorthiamid is prepared by the action of hydrogen sulphide on 2,6-dichlorobenzonitrile in the presence of pyridine.

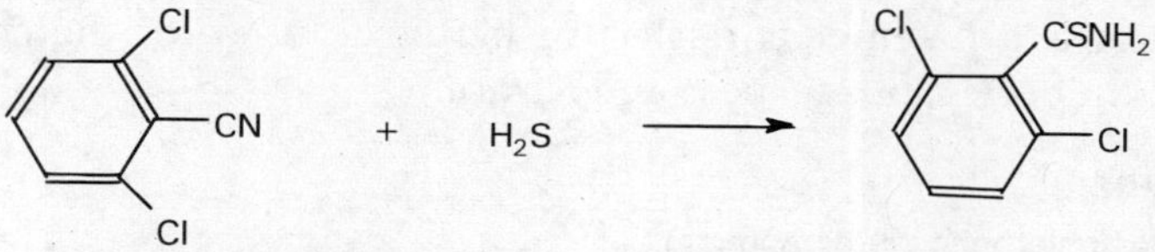

It is used for the control of weeds in rice and in orchards.

### Ioxynil

(3,5-Diiodo-4-hydroxybenzonitrile)

Ioxynil is produced by the direct iodination of 4-hydroxybenzaldehyde followed by conversion to the oxime and subsequent dehydration to nitrile.

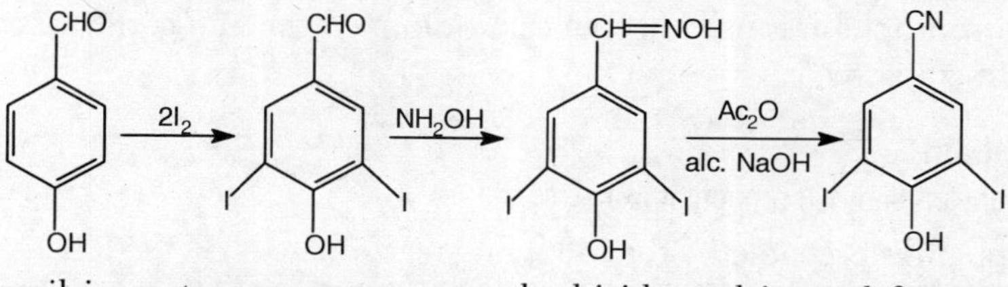

Ioxynil is post emergence contact herbicide and is used for control of broad leaved weeds in cereals and monocot crops.

### Bromoxynil

(3,5-dibromo-4-hydroxybenzonitrile)

Bromoxynil is produced by reacting 4-hydroxybenzonitrile with sodium hypobromite.

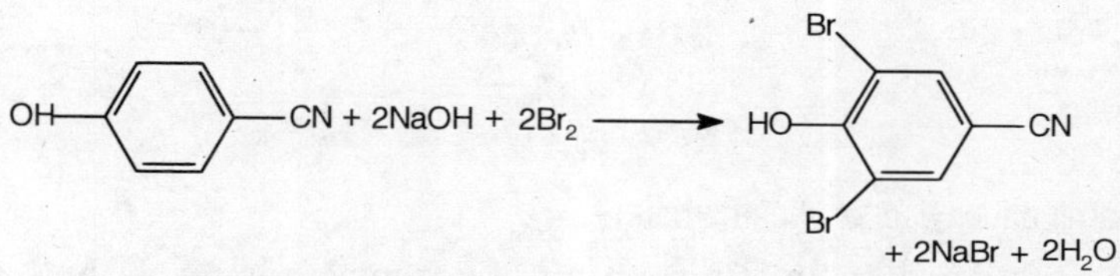

Bromoxynil is a post emergence contact herbicide and is used for control of broad leaved weeds in cereals and grass crops.

## CARBAMATE HERBICIDES

Carbamates are esters of carbamic acid ($NH_2COOH$). The alkyl esters of N-aryl carbamic acids are powerful herbicides.

### Propham

(Isopropyl-N-phenylcarbamate)

Propham is produced by the reaction of phenyl isocyanate with isopropyl alcohol

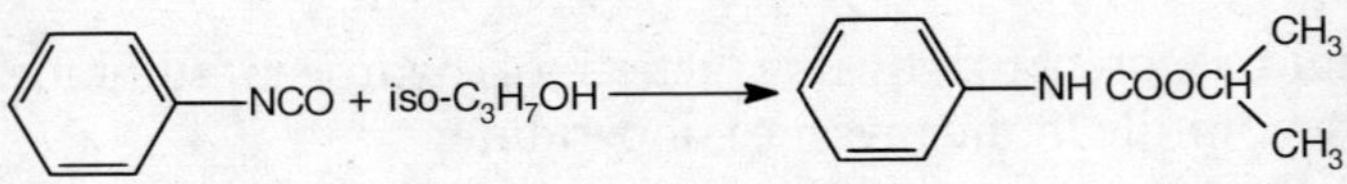

Propham is a preemergent selective herbicide used for the control of grass weeds in beet, cabbage, peas and lettuce.

### Chloropropham

(Isopropyl-N-3(chlorophenyl) carbamate)

Chloropropham is prepared by reacting m-chlorophenylisocyanate with isopropyl alcohol

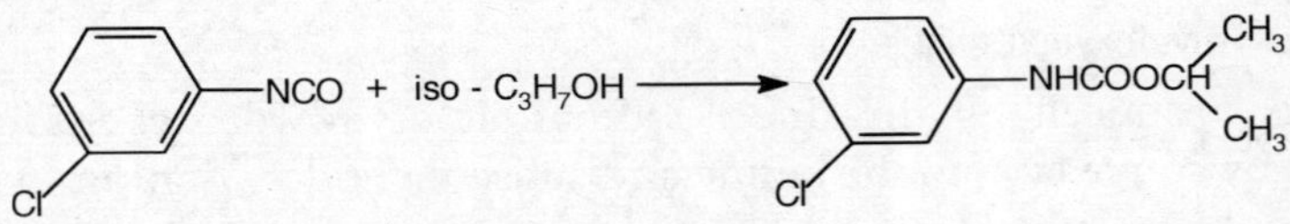

Chloropropham is used in soybean, onion, cotton and garlic for control of several grass weeds.

### Proximpham

(O-(N-phenylcarbamoyl) propanon oxime)

Proximpham is prepared by the following reaction

Proximpham is a selective soil herbicide. It is used for the control of weeds in beet, onion and vegetables.

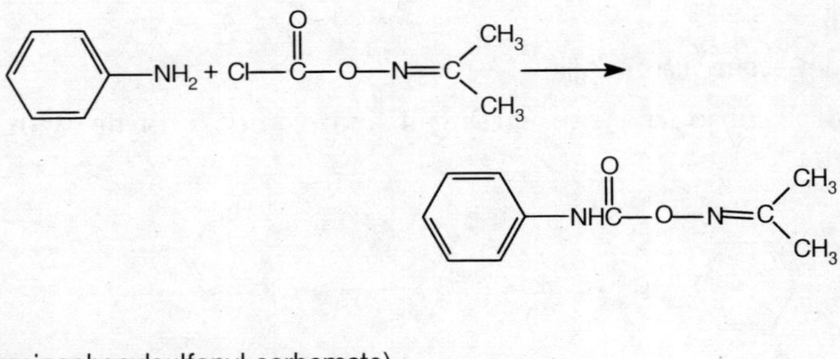

### Asulam

(methyl 4-aminophenylsulfonyl carbamate)

Asulam is prepared by treating 4-aminobenzenesulfonamide and methyl chloroformate in the presence of acid binding agent

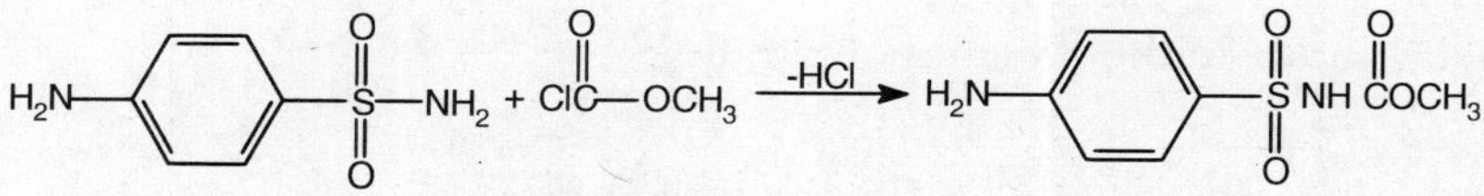

It is effective against weed grasses. The acute oral $LD_{50}$ of its potassium salt for rats is 2000 mg/kg.

## Phenmedipham

(3-methoxycarbonylaminophenyl-N (3'-methylphenyl) carbamate)

Phenmedipham is prepared by treating m-amino phenol with methylchloroformate to give 3-methoxycarbonyl aminophenol which on reacting with m-tolyl isocyanate gives phenmedipham.

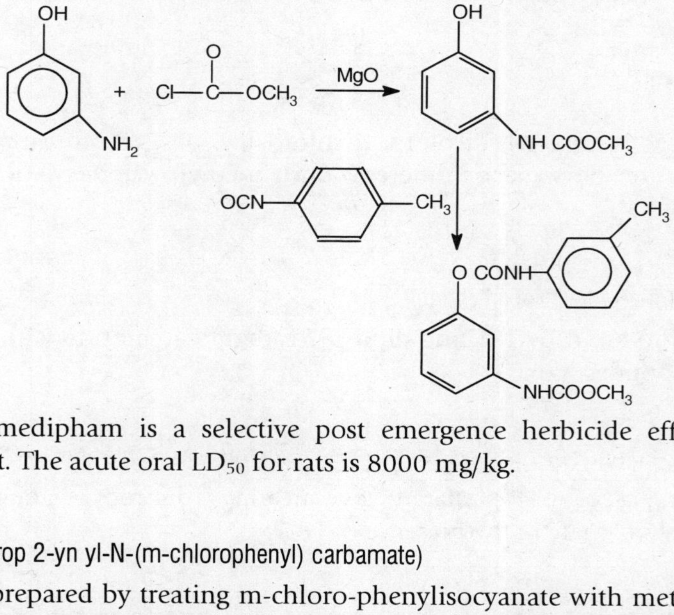

Phenmedipham is a selective post emergence herbicide effective in sugar beet. The acute oral $LD_{50}$ for rats is 8000 mg/kg.

## Bi-PC

(1-Methylprop 2-yn yl-N-(m-chlorophenyl) carbamate)

Bi-PC is prepared by treating m-chloro-phenylisocyanate with methyl ethyl carbinol or methyl propynl chlorocarbonate with m-chloroaniline in the presence of base.

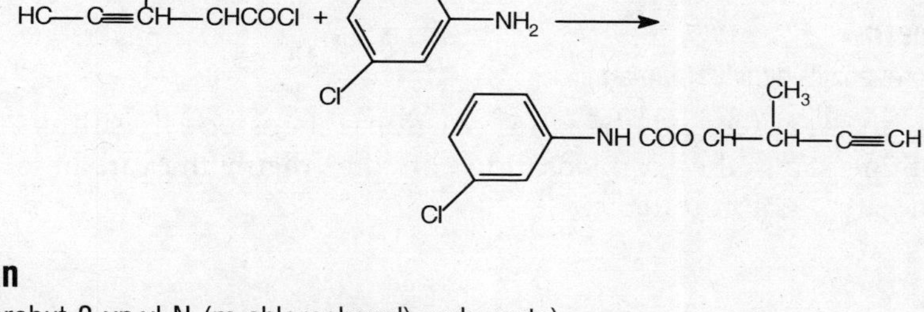

## Barban

(4-chlorobut-2-yn yl-N-(m-chlorophenyl) carbamate)

It is prepared by reacting m-chlorophenyl isocyanate with butyndiol and thionyl chloride in the presence of pyridine

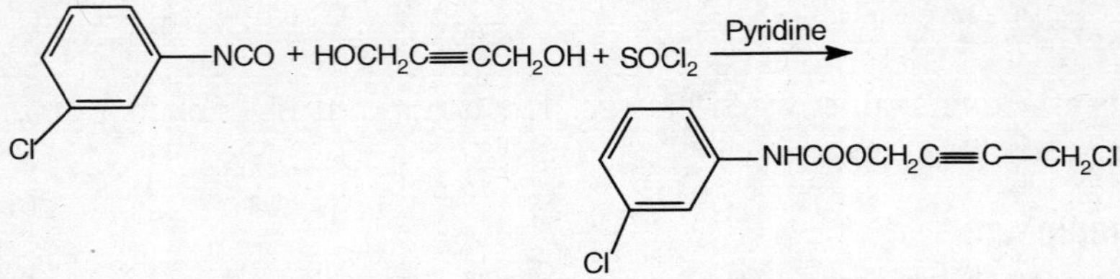

Barban is used to control wild oats.

## THIOCARBAMATES

This group of herbicides is characterised by the general formula

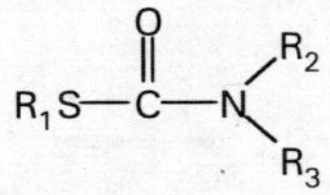

where $R_1$ is an alkyl, dichloro, trichloroalkyl or 4-chlorobenzyl group, $R_2$ and $R_3$ are the same or different alkyl, methyl cyclohexyl or azepino groups.

### Eptam

(S-ethyl N,N-di-n-propylthiocarbamate)

Eptam is produced by reacting dipropyl carbamoyl chloride with sodium ethyl mercoptides.

$$(n\ C_3H_7)_2\ NCOCl + C_2H_5SH + NaOH \longrightarrow (n\ C_3H_7)_2\ N\overset{O}{\overset{\|}{C}}SC_2H_5 + NaCl + H_2O$$

It can also be prepared by treating di-propyl amine with thiochlorocarbamate in the presence of HCl.

$$(n\text{-}C_3H_7)_2NH + C_2H_5\ SCOCl + NaOH \longrightarrow (n\text{-}C_3H_7)_2\ N\overset{O}{\overset{\|}{C}}S\ C_2H_5 + NaCl + H_2O$$

It is used for control of weeds in potato, alfalfa, beans, carrot, corn, sugarbeet, sunflower, tobacco and number of vegetable crops.

### Benthiocarb

(S-(4-chlorobenzyl) dithiocarbamate)

Benthiocarb is prepared by the reaction of carbonyl sulfide with diethylamine and the condensation of the diethylthiocarbamic acid obtained with 4-chlorotoluene.

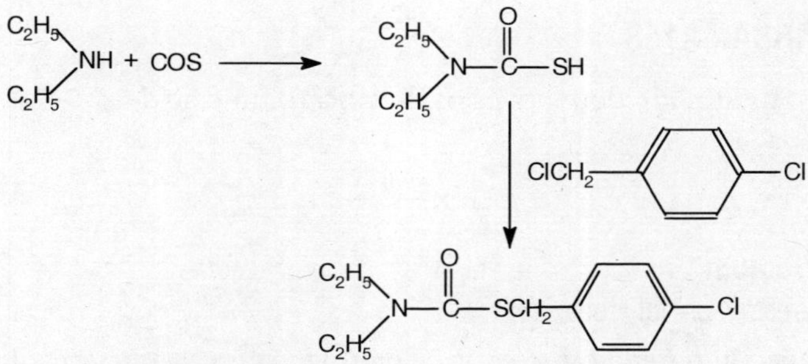

Benthiocarb is a selective herbicide in rice but is also used for weed control in cotton, soybean, potatoes, peanut, beet and beans.

## Diallate

(S-2,3-dichloroallyl diisopropylthiocarbamate)

Diallate is prepared by treating diisopropylamine, carbonyl sulfide and 1,2,3-trichloropropene in the presence of base.

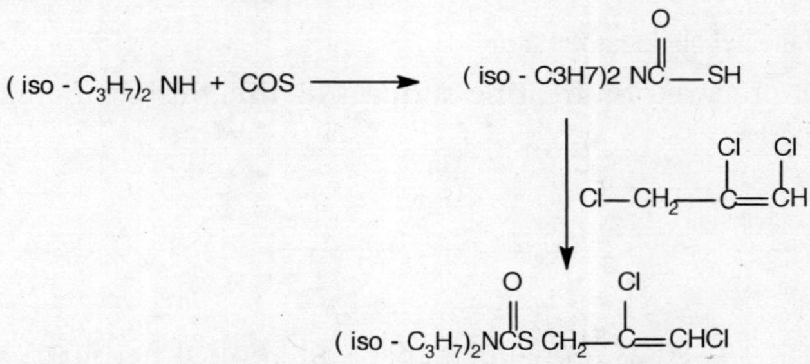

Diallate is used for the control of wild oat and block grass.

## Triallate

(S-2,3,3-trichloroallyl diisopropyl thiocarbamate)

Triallate is prepared by treating diisopropylamine, carbonyl sulfide and 1,1,2,3-tetrachloroprop-1-ene in the presence of a base.

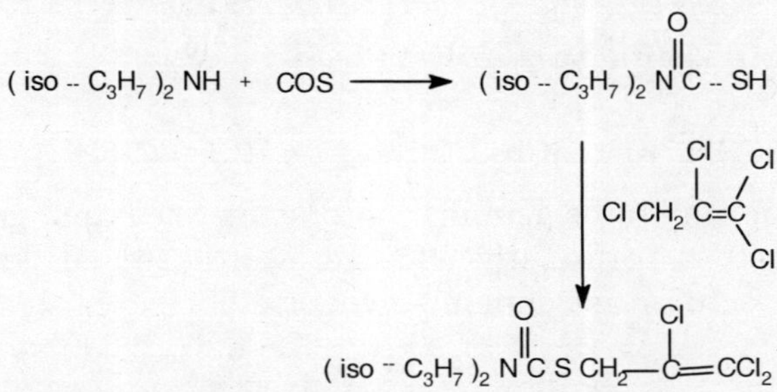

Triallate is selective in cereals, wheat and barley.

## DITHIOCARBAMATES

Dithiocarbamates are derivatives of dithiocarbamic acid

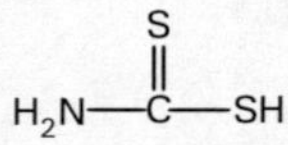

### Metham-Sodium

(Sodium-N-methyl dithiocarbamate)

Metham is prepared by treating methylamine, carbondisulphide and sodium hydroxide.

$$CH_3NH_2 + CS_2 + NaOH + H_2O \longrightarrow CH_3NH\overset{S}{\overset{\|}{C}}SNa + 2H_2O$$

During the decomposition of metham, methylisothiocyanate is formed and this exerts the biological action.

### Sulfallate

(2-chloroallyl diethyl dithiocarbamate)

Sulfallate is prepared by treating sodium diethyldithiocarbamate with 2,3-dichloropropene.

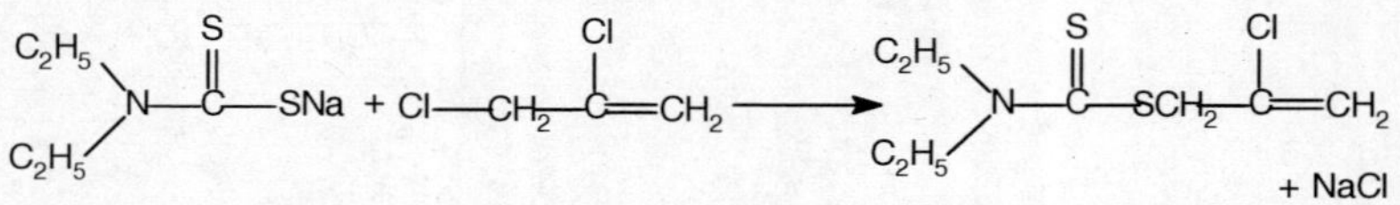

Sulfallate is used as a preemergence soil applied herbicide against grass weeds in vegetable crops such as brassica, celery, spinach, beans, corn and in ornamentals.

## AMIDES

### Allidochlor

(N,N-diallyl-2-chloroacetamide)

Allidochlor is prepared by the raction of diallylamine and chloroacetyl chloride.

$$(CH_2 = CHCH_2)_2NH + ClCH_2\ COCl \xrightarrow{\text{base}} (CH_2 = CHCH_2)_2NCOCH_2Cl$$

Allidochlor inhibit the germination of grass weeds and growth of roots and shoots and is particularly effective against annual grasses in corn, sorghum, beans, sugarcane and ornamentals.

### Propachlor

(2-chloro-N-isopropylacetanilide)

Propachlor is prepared by treating N-isopropylaniline with chloroacetylchloride in the presence of sodium hydroxide.

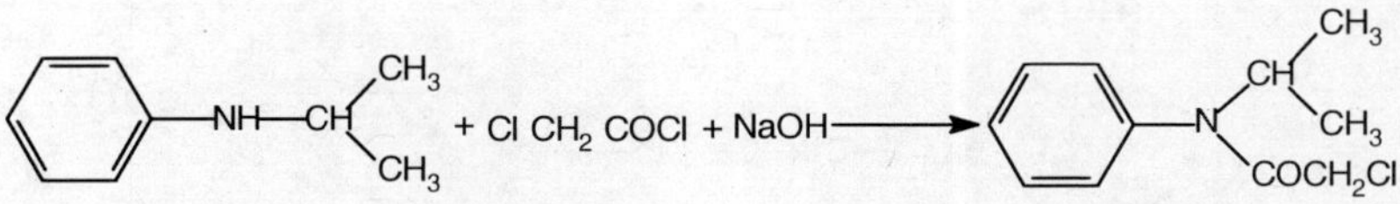

Propachlor is a preemergence selective herbicide effective against annual grass weeds and several broad-leaved weeds in maize, cotton, sugarcane, soybean and several vegetable crops. The acute oral $LD_{50}$ of propachlor to rats is 710 mg/kg.

## Prynachlor

(2-chloro-N-(1-methyl-2-propynyl) acetanilide)

Prynachlor is prepared by treating butynyl chloride with aniline to give N-(1-methyl-2-propynyl) aniline which is then reacted with chloroacetyl chloride to give prynachlor.

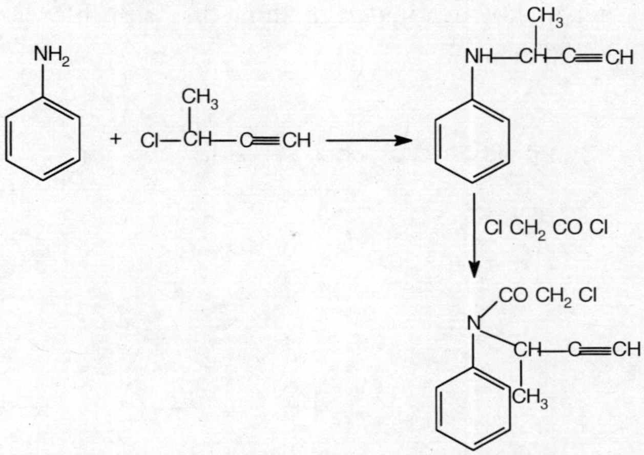

Prynachlor is a preemergence selective herbicide effective in maize, soybean and onion against annual grass weeds and several broad leaved weeds.

## Alachlor

(2-chloro-2',6'-diethyl-N-methoxymethylacetanilide)

Alachlor is prepared from azomethine substituted at the phenyl group by reaction first with chloroacetyl chloride and then with methyl alcohol.

Alachlor is a selective preemergence soil herbicide and is effective against annual grass weeds and several broad leaved weeds in maize, soybean, peanut, cotton and fruit trees.

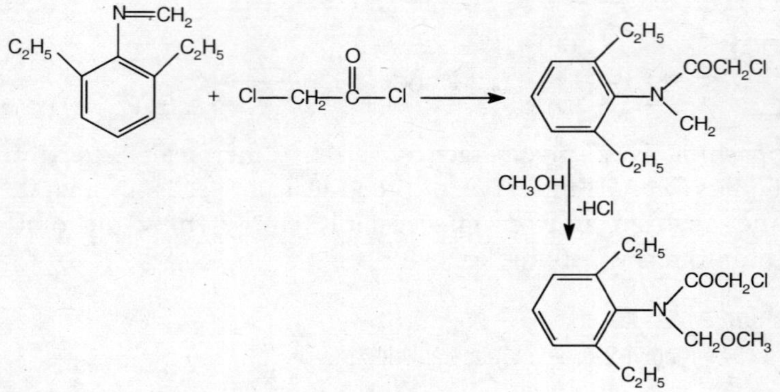

## Butachlor

(2-chloro-2',6'-diethyl-N-butoxymethyl acetanilide)

Butachlor is prepared by treating azomethine first with chloroacetylchloride and then with butyl alcohol

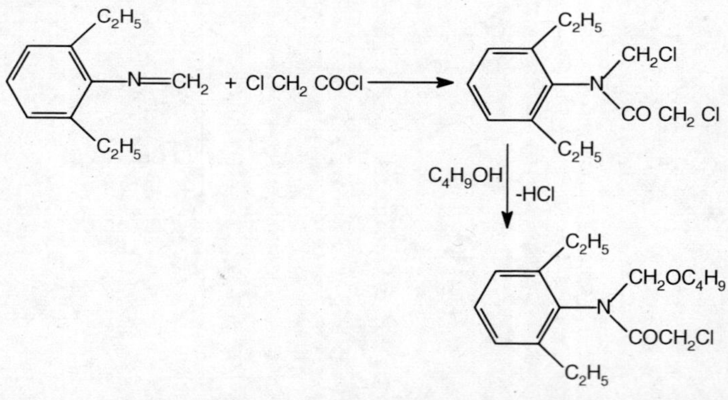

Butachlor controls most of the annual grasses and some broad leaved weeds in transplant and direct seed rice. It is applied as pre-emergence in EC formulations and as early post emergence in the form of granules. Its acute oral $LD_{50}$ for rats is 1740 mg/kg.

## Diphenamid

(N,N-dimethyl-2,2-diphenyl acetamide)

Diphenamid is used as pre-emergence soil herbicide against annual grasses and certain broad leaved weeds. It is prepared by the condensation of sodium salt of diphenyl methane and dimethyl carbamoyl chloride.

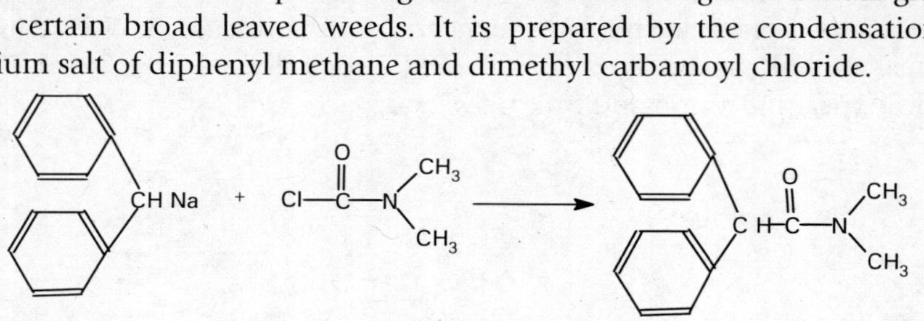

### Propanil

(3'4'-dichlorophenyl propionanilide)

Propanil is extensively used for selective post-emergence weed control in rice fields. The rice is resistant because it can deactivate the chemical by hydrolysis to 3,4-dichloroaniline. It can be prepared by treating 3,4-dichloroaniline with propionyl chloride in the presence of triethylamine.

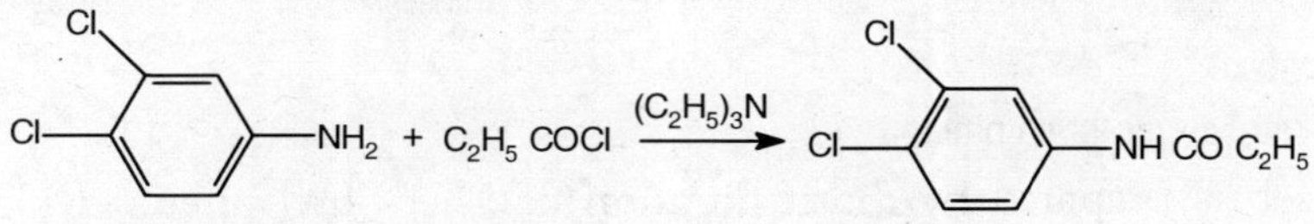


## BIPYRIDYLIUM HERBICIDE

### Paraquat
### (1,1, Dimethyl-4,4, dipyridybium chloride).

It is synthesied by treating 4,4′-dipyridyl with methyl chloride at high temperature and pressure.

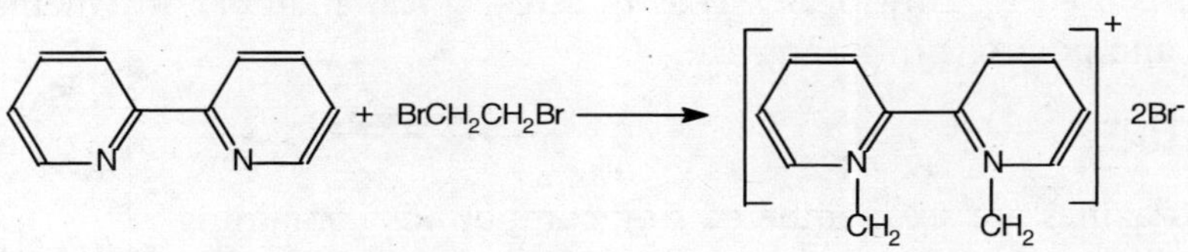


### Diquat

(1,1, ethylene-2,3-dipyridybium bromide)

It is prepared by reacting 2,2′ dipyridyl with 1,2, dibromoethane

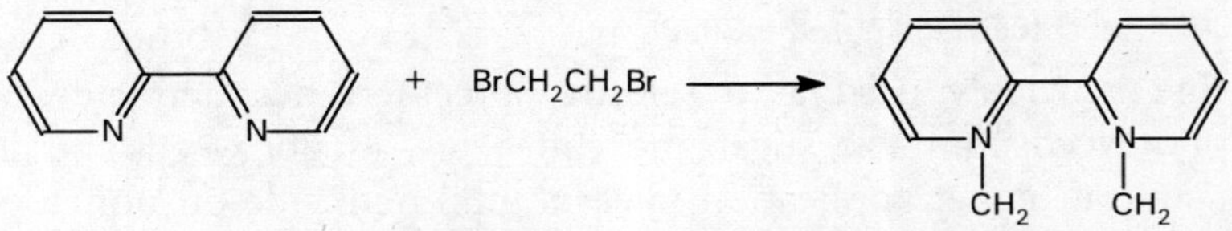


Diquat is used for general aquatic weed control as noncropweed killer and for defoliation. Paraquat is used for stubble cleaning and for total weed control in non crop areas.

## DINITROPHENOLS

### DNOC

(2,4-Dinitro-6-methylphenol)

DNOC is produced by the direct nitration of o-cresol with a nitrating mixture at low temperature.

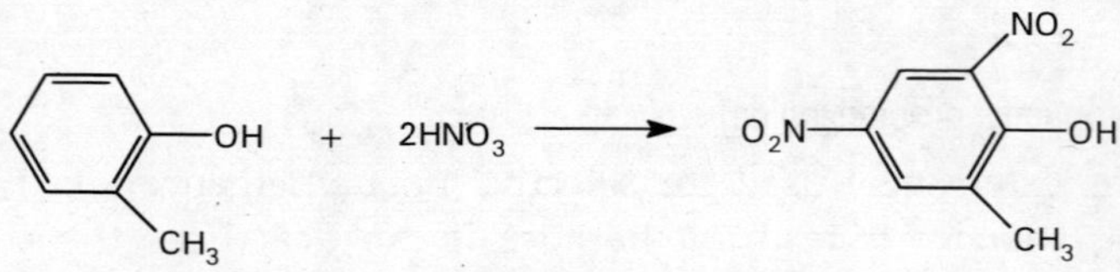

It is used for control of annual broad-leaf weeds in cereals, dessicant for potatoes and leguminous seed crops.

## Dinoseb

(2,4-dinitro-6-sec-butyl phenol)

Dinoseb is prepared by direct nitration of 2-sec-butyl phenol with nitric acid in the presence of sulphuric acid.

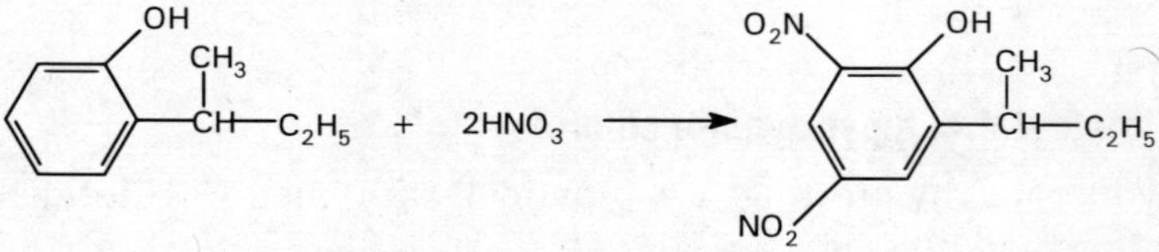

It is used for control of weeds in cereals, pea, alfalfa (post emergence), bean and potato (pre-emergence).

## TRIAZINES

The triazines are six membered ring compounds containing three heterocyclic nitrogens (the prefix tri means three and azine means nitrogen containing). The fundamental triazine nucleus is shown as:

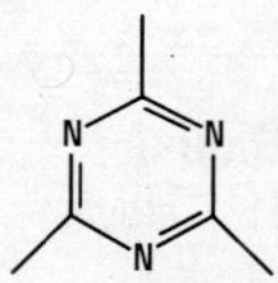

## Simazine

(2-chloro-4,6-bis (ethyl amino) -s-triazine)

Simazine is widely used as a selective preemergence herbicide for the control of weeds in maize, sugarcane, coffee, tea, aquatic weeds and algae in ponds and at higher application rates as total herbicide on noncrop land. Simazine is prepared by reaction with cyanuric chloride with ethylamine and NaOH in aqueous medium or in organic solvents.

$$Cl_2 + NaCN \longrightarrow CNCl + NaCl$$

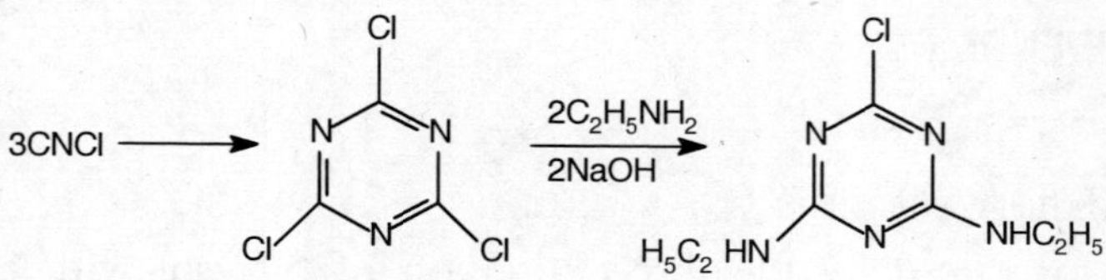

## Atrazine

(2-chloro-4-(ethylamino)-6-(isopropylamino)-s-triazine)

Atrazine is a selective preemergence herbicide for the control of broad leaf weeds in maize, sorghum, sugarcane, pine apple, and orchards. It is also used as nonselective herbicide in non crop areas. It is prepared by treating cyanuric chloride with ethyl amine and NaOH which produces 2,4-dichloro-6-ethylamine-s-triazine and by further reaction with a molecule of isopropylamine and NaOH gives atrazine.

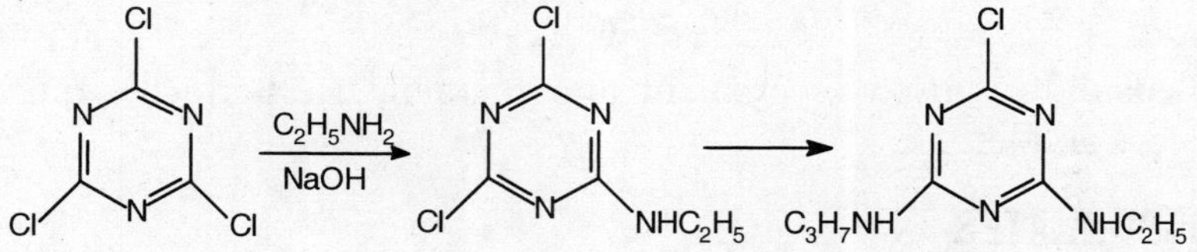

### Prometryne

(2,4 bis (isopropylamine)-6-(methylthio)-s-triazine)

Prometryne is used for control of weeds in cotton. It is prepared by reacting 2-chloro-4,6-bis (isopropyl ammino)-S-triazine with sodium mercoptide.

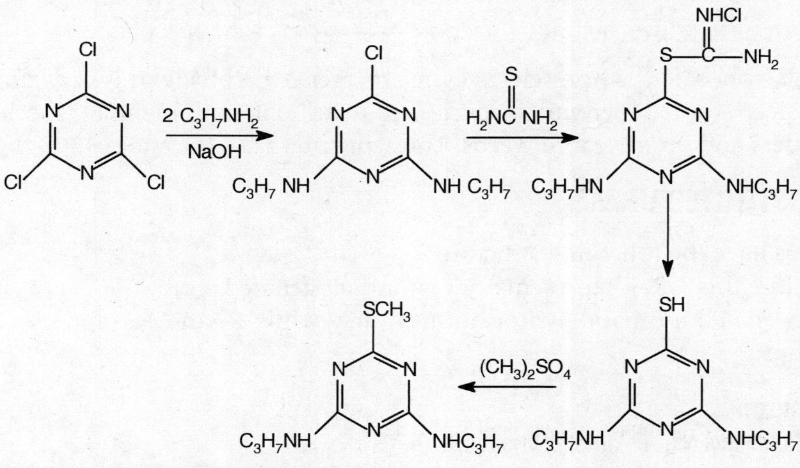

## ORGANOPHOSPHOROUS COMPOUND

A wide range of different types of organophosphorous compounds show some herbicide properties and the major groups are phosphorothiolates, phosphoroamidates and phosphonates.

## PHOSPHOROTHIOLATES

### Bensulide

(O,O-disopropyl-S-(2-phenyl sulfonylamino) ethyl phosphorodithioate)

Bensulide is prepared by the condensation of sodium O,O-diiso-propylphosphorodithioate with N-(2-chloroethyl) benzene sulfonamide

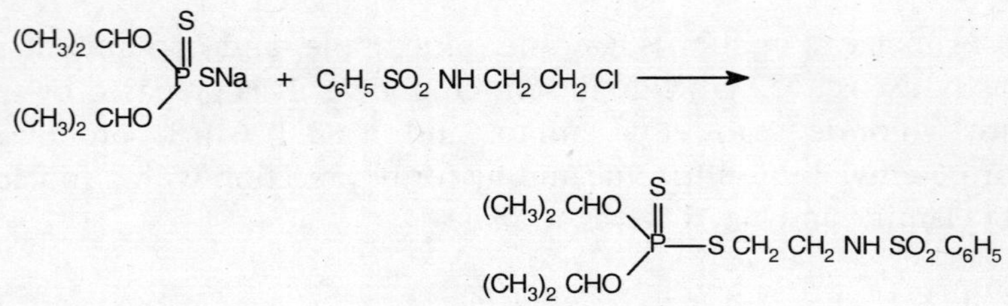

Bensulide is used as preplant and premergence herbicide on cotton, brassicos and lettuce.

## PHOSPHONATES

### Glyphosate

(N-phosphonomethyl glycine)

It is obtained from glycine and chloromethyl phosphonic acid.

$$(HO)_2\overset{O}{\overset{\|}{P}}CH_2\,Cl + NH_2\,CH_2\,COOH \xrightarrow{OH^-} (HO)_2\,\overset{O}{\overset{\|}{P}}\,CH_2\,NH\,CH_2\,COOH$$

Glyphosate is a non selective pre emergence herbicide. It is very effective on deep rooted perennial species, and annual and biennial species of grassess and broad leaved weeds. The acute oral $LD_{50}$ for rats is 4320 mg/kg.

## SUBSTITUTED UREAS

Ureas have the following structure

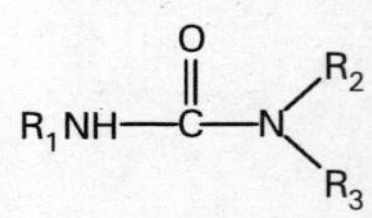

In this $R_1$ represents a non-halogenated or halogenated aromatic hydrocarbon moiety while $R_2$ and $R_3$ represent alkyl groups.

### Monuron

(N,N-dimethyl N'(p-chlorophenyl) urea)

Monuron the first urea herbicide developed for commercial use, introduced as a pre-emergence, broad spectrum herbicide for control of all weeds in non-crop land areas. When applied at low rates, monuron show selectivity for use in cotton, sugarcane, pineapple and citrus. It is prepared as follow:

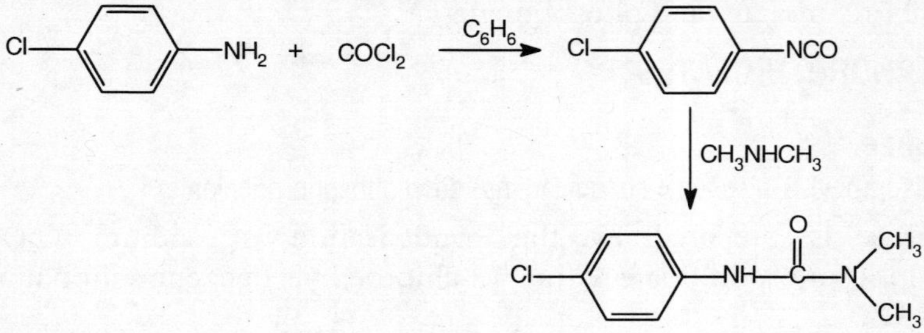

## Linuron

(3(3,4-dichlorophenyl)-1-methoxy-1-methyl urea)

It is prepared as follows:

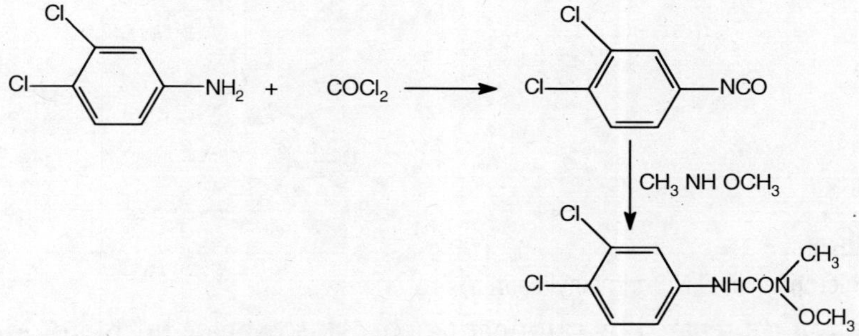

It is used as pre emergence for slective control of germinating and newly established broadleaf weeds and grasses in soybean, cotton, corn, sorghum, wheat and potato.

## Diuron

(3-(3,4-dichlorophenyl) 1,1-dimethyl urea)

It is preemergence selective herbicide for controlling emerging grass and broadleaf weed seedlings in cotton, tea, coffee, grapes, apples, pears and citrus etc. It is prepared as:

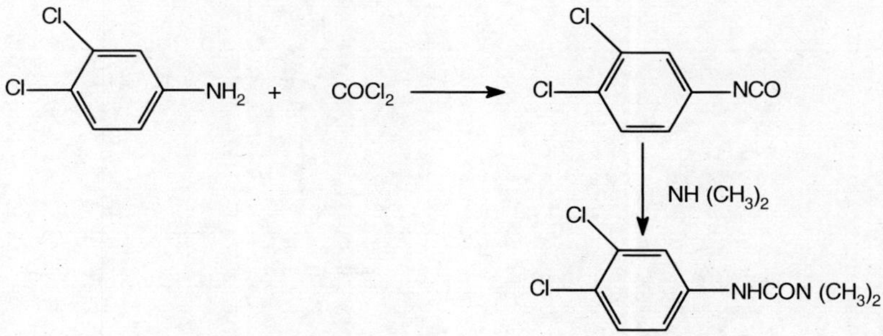

## Neburon

(N-Butyl N'-(3,4-dichlorophenyl) N-methyl urea)

It is used as selective root absorbed herbicide for control of annual broad leaf and grass weeds. It is prepared as follows:

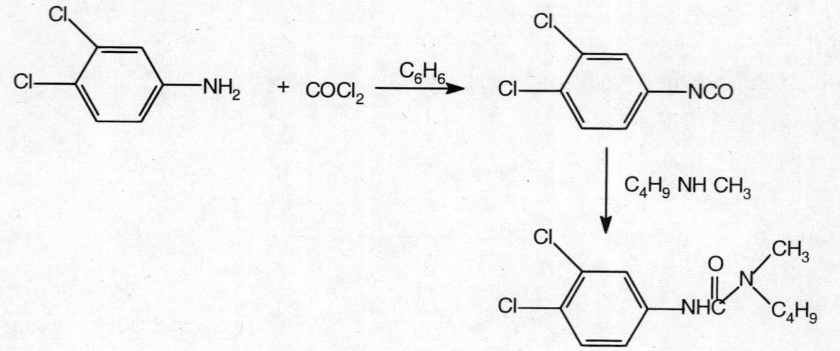


## Trimeturon

(N-(4-chlorophenyl) O,N,3-trimethylisourea)

Trimeturon is prepared by chlorinating phenyl isocyanate in chloroform to give p-chlorophenyl carbonim idoyl dichloride, which is reacted with dimethylamine to give p-chlorophenylimino chlorocarbonic acid dimethyl amide. This is treated with sodium methoxide solution in benzene to yield trimeturon.

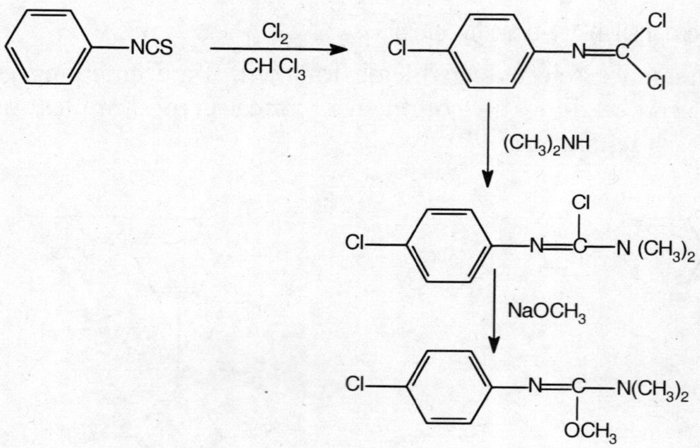


## Meturin

(1-phenyl-1-hydroxy-3 methyl urea)

Meturin is prepared by reaction of phenylhydroxylamine with methyl isocyanate.

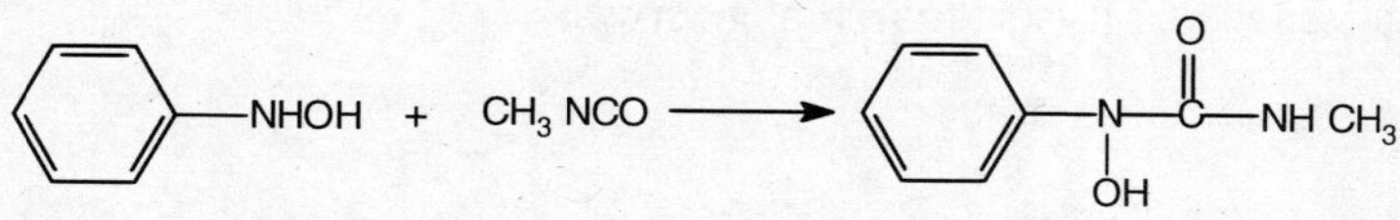


Meturin is used as a pre emergence herbicide for the control of annual weeds in cotton and potatoes. Its acute oral $LD_{50}$ for rats is 5000 mg/kg.

## Dichloral Urea

(N,N-di-(2,2,2-trichloro-1-hydroxyethyl) urea)

Dichlorol urea is prepared by the condensation of chloral and urea in the presence of HC1

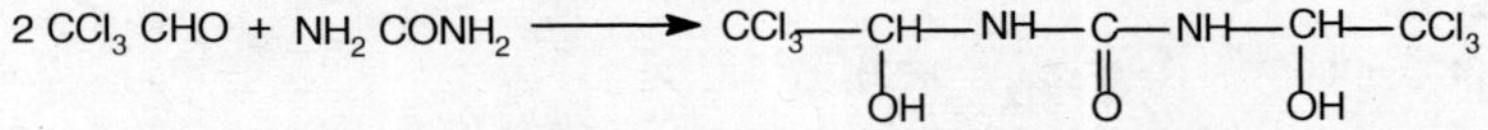


Dichloral urea is a soil herbicide for pre emergence treatment in sugar beet, cotton, potato and carrots.

## Cycluron

(3-cyclooctyl-1,1 dimethyl urea)

Cycluron is prepared by treating cyclooctyl isocyanate with dimethyl amine.

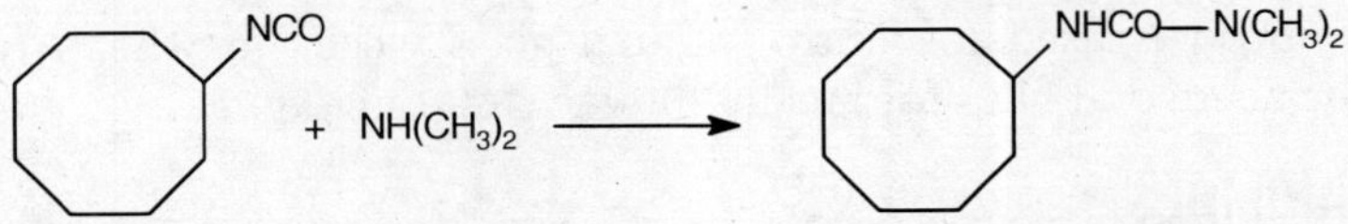


Cycluron is used for pre emergence weed control in sugar beet and spinach.

## Noruron

(3-(hexahydro-4,7-methanoindan-5-yl) 1,1-dimethyl urea)

Noruron is prepared by the condensation of tetrahydrodicyclopentadienyl isocyanate with dimethylamine

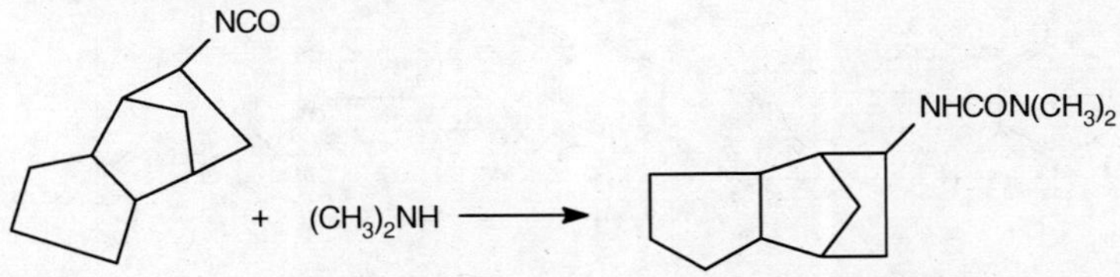


Noruron is a selective pre emergence herbicide on cotton, sugarcane soybeans and potato.

## Monolinuron

(3-(4-chlorophenyl)-1-methoxy-1-methyl urea)

Monolinuron is prepared by treating aryl isocyanate and hydroxylamine to give 3-aryl-1-hydroxy urea, which is then treated with dimethyl sulphate.

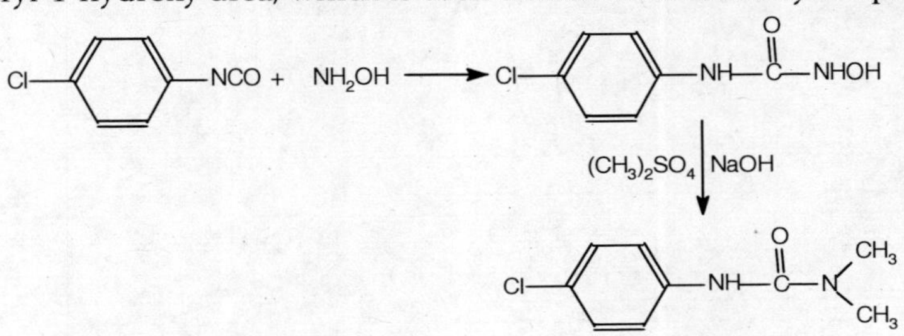

It is used as a pre emergence herbicide for the control of weeds in potatoes, medicinal plants, vineyards and orchards. Its acute oral $LD_{50}$ for rats is 2250 mg/kg.

## Isoproturon

(3,(4-isopropylphenyl)1,1-dimethyl urea)

Isoproturon is prepared as:

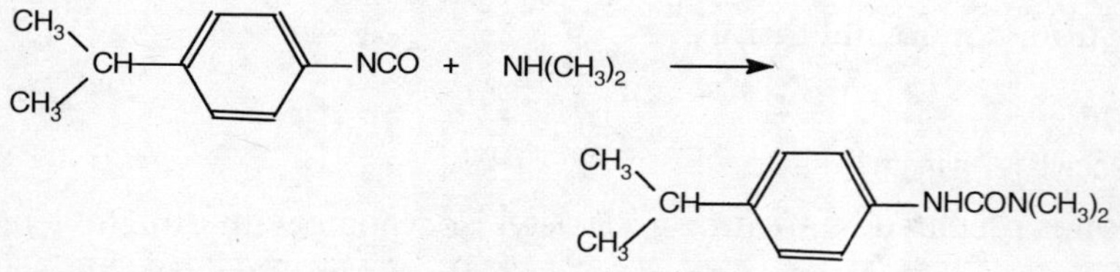


Isoproturon is used to control annual grass weeds in wheat, rye and barley. The acute oral $LD_{50}$ for rats in 2000 to 3600 mg/kg.

# DINITROANILINES

## Nitralin

(2,6-dinitro 4-methyl sulphonyl N,N-dipropylaniline)

Nitraline is prepared by nitration of 4-chlorophenyl methyl sulphone and coupling of 4-chloro-3,5-dinitrophenyl methyl sulfone obtained with dipropylamine.

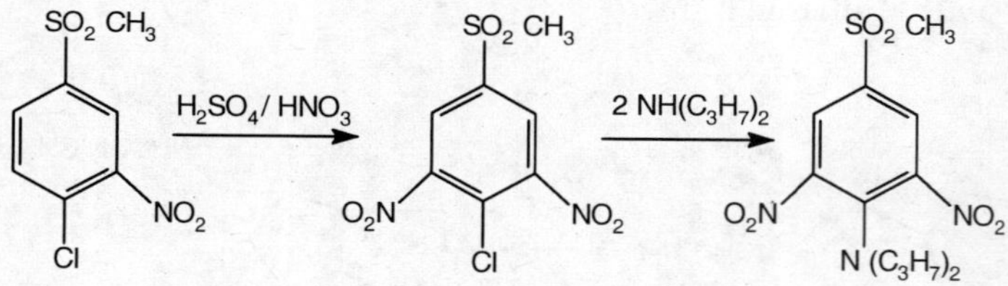


## Dipropalin

(N,N-di-n-propyl-2,6-dinitro-4-methyl aniline)

Dipropalin is prepared by converting p-cresol into dinitrocresol which is then treated with thionyl chloride. The 4-chloro-3,5-dinitrotoluene so obtained is reacted with di-n-propylamine to give dipropalin.

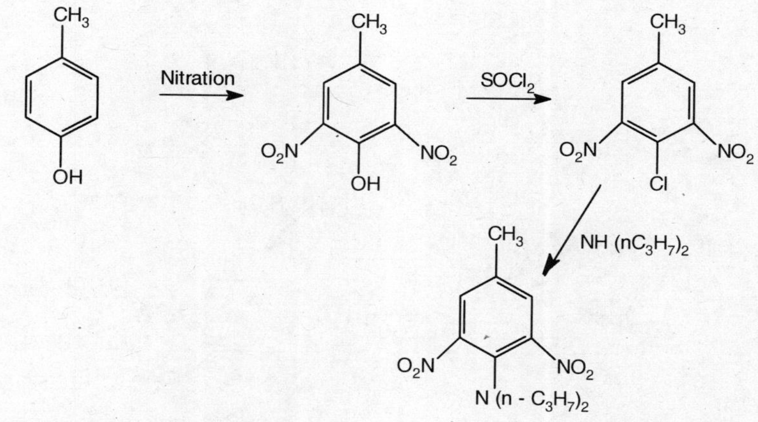


## Butralin

(N-sec-butyl-4-t-butyl 2,6-dinitroaniline)

Butralin is prepared by nitration of 4-t-butylphenol with a mixrture of nitric acid and sulphuric acid to give 4-t-butyl-2,6-dinitrophenol. The dinitrophenol is reacted with phosphorous oxychloride to give 1-t-butyl-4-chloro-2-dinitrobenzene. The reaction of the later with sec-butylamine give butralin.

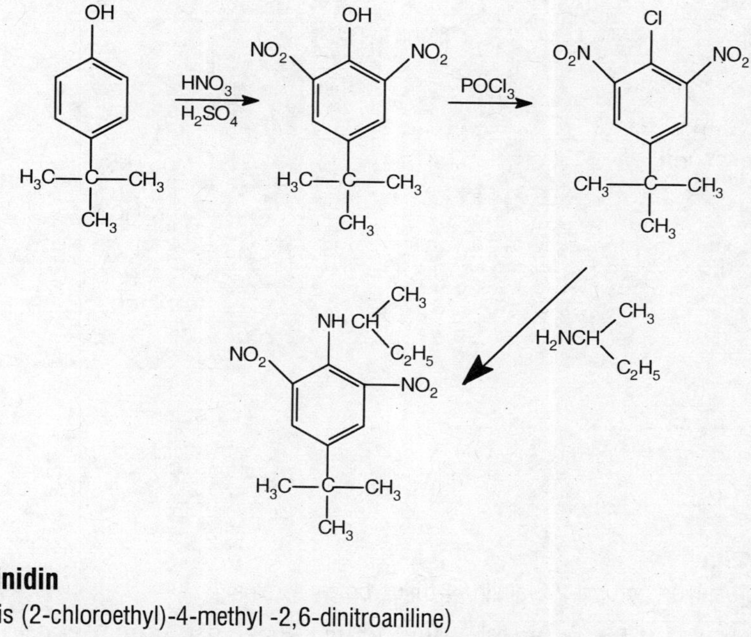


## Chlorinidin

(N,N-bis (2-chloroethyl)-4-methyl -2,6-dinitroaniline)

Chlorinidin is prepared by treating 4-chloro-3,5-dinitro toluene with diethanolamine and the resulting compound is treated with thionyl chloride.

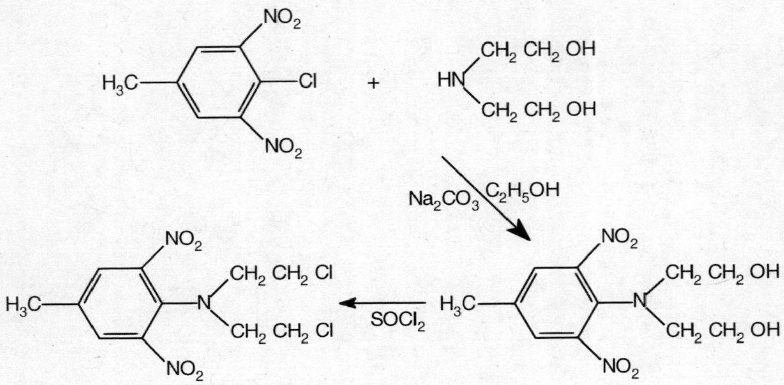

**Isopropalin**

(4-isopropyl-2,6-dinitro-N,N-dipropylaniline)

Isopropalin is prepared by nitration of cumene to give 2,4-dinitrocumene which is then converted with phosphorous oxychloride into 1-chloro-2,6-dinitro-4-isopropylbenzene, the later giving isopropalin with di-n-propylamine. The later giving isopropalin with di-n-propylamine.

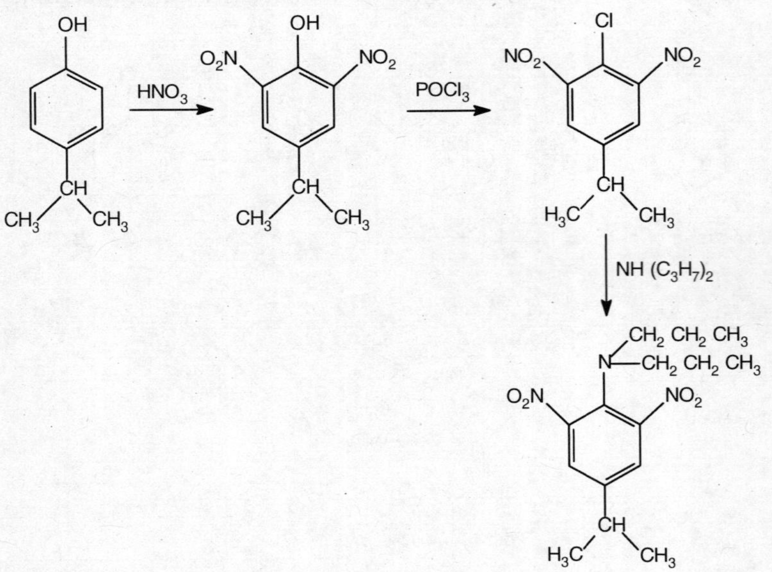

**Fluchloralin**

(N-2-chloroethyl)-N-propyl-2,6-dinitro-4-trifluoromethylamine)

Fluchloralin is used as pre planting herbicide against annual grasses and broad leaf weeds in cotton and soybean.

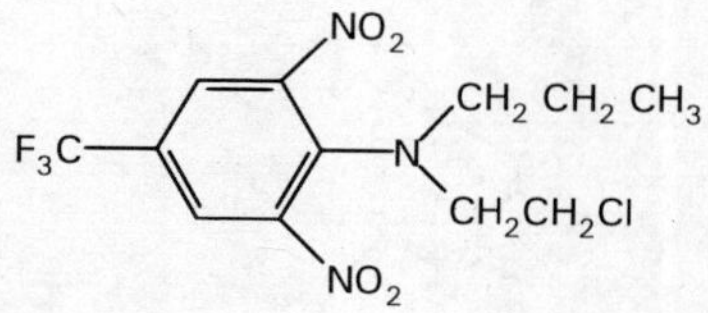

### Pendimethalin

(Stomp, N-(1-ethyl propyl) 3,4-dimethyl-2,6-dinitro aniline)

Pendemethalin is the first N-monoalkyl 2,6-dinitroaniline and is mainly used for control of grasses in maize, cereals, rice, cotton and soybeans. Its $LD_{50}$ value to rats is 1050 mg/kg.

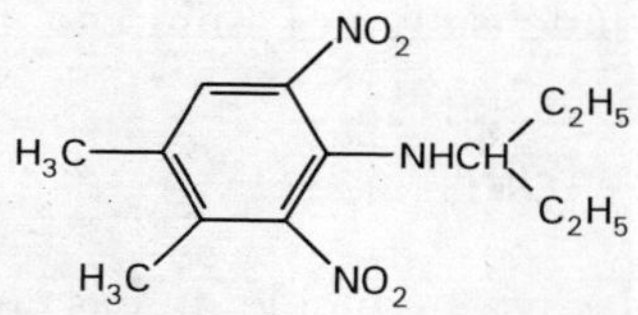

The main uses of dinotroaniline herbicides is the control of annual grass weeds. Some annual dicotyledonous weeds are also sensitive to dintroaniline herbicides. All herbicides of this group are fairly volatile and hence are sprayed to moist soil surface.

## PYRIDINES

### Pyrichlor

(2,3,5-trichloro-4-pyridinol)

Pyrichlor is prepared by the alkaline saponification of 2,3,4,5-tetrachloropyridine

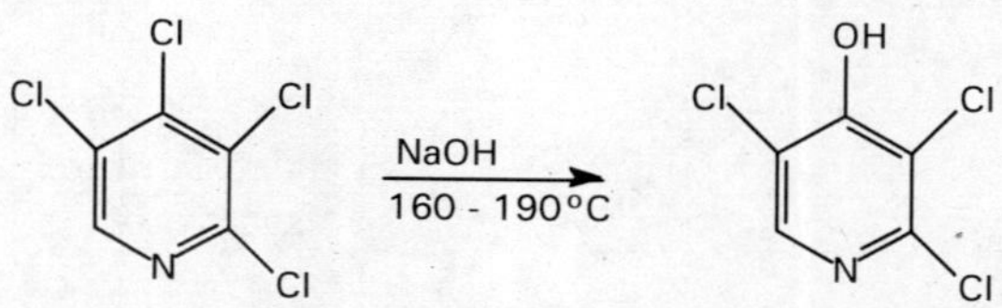

Pyrichlor is used as preemergence herbicide for the control of annual grass weeds and broad leaved weeds. It is also used for selective weed control in sugar cane, banana and rubber plantations.

Other important herbicide of the pyridine groups are:

- Picloram (4-amino-3,5,6-trichloropicolinic acid)
- Trichlopyr (3,5,6-trichloro-2-pyridinyloxyacetic acid)
- DPA (3,6-dichloropicolinic acid)

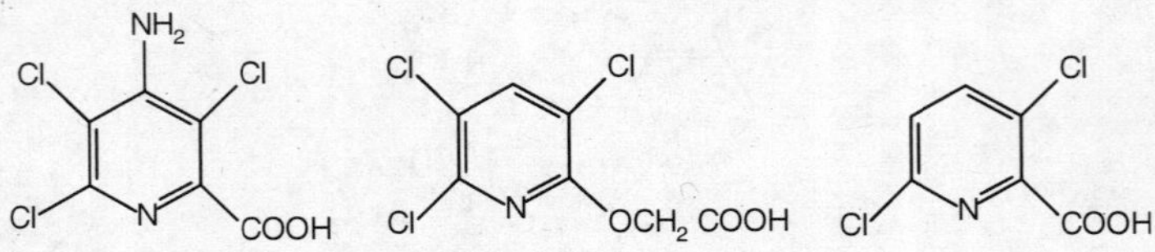


Picloram Triclopyr DPA

## AZOLES

### Amitrole
### (3-amino-1,2,4-triazole)

Amitrole is prepared by condensation of aminoguanidine and formic acid

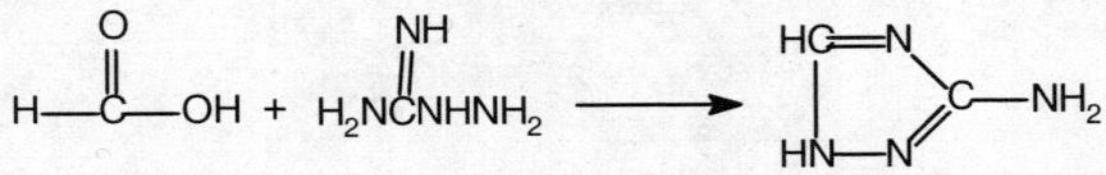


Amitrole is a non-selective herbicide and is used for control of deep rooted weeds.

### Methazole

(2-(3,4-dichlorophenyl)-4-methyl-1,2,4-oxadiAzolidine-3,5-dione)

Methazole is prepared by the reaction of N-(3,4-dichlorophenyl)-hydroxylamine with methyl isocyanate to give 1-(3,4-dichlorophenyl)-1-hydroxy-3-methyl urea which is cyclised with methyl chloroformate to give methazole

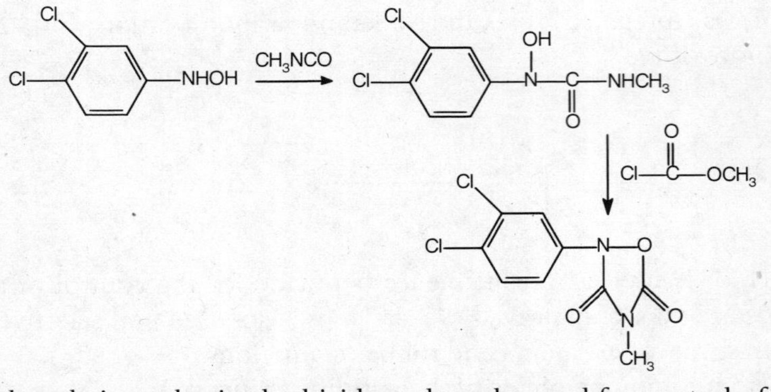


Methazole is a selective herbicide and can be used for control of certain grasses and many broad leaved weeds.

### Oxadiazon

(5-t-butyl-3-(2,4-dichloro-5-isopropoxyphenyl)1,3,4-oxadiazol-2-one)

Oxadiazon is obtaioned by condensing trimethyl acetylchloride with 2,4-dichloro-5-isopropoxyphenyl hydrazine and the product is cyclised with phosgene to give oxadiazon.

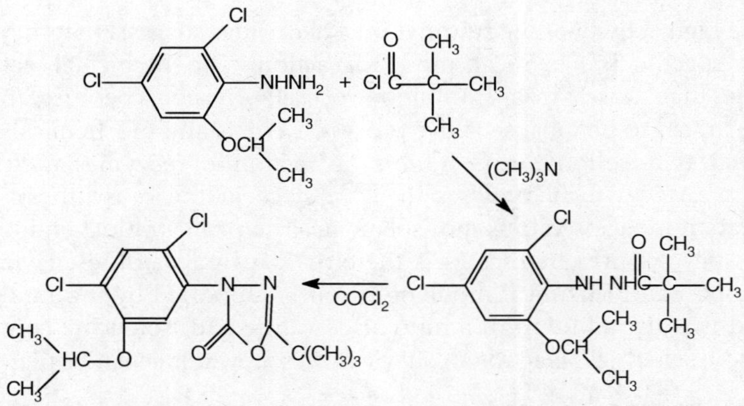

Oxadiazon is selective herbicide, effective against mono- and dicotyledonous weeds in rice, orchards and vine yards. The acute oral $LD_{50}$ for rat is 8000 mg/kg.

## SULFONYL UREA HERBICIDES

Sulfonylurea herbicides are new group of selective herbicides and are used for the control of broad-leaved weeds. These herbicides are applied at very low dosage of 10-40 g $ha^{-1}$. Important sulfonyl urea herbicides are:

- chlorsulfuron (1-(2-chlorophenylsulfonyl)-3-(4-methoxy-6-methyl-1,3,5-triazin-2-yl) urea
- metasulfuron-methyl 2-(3-(4-methoxy-6 methyl-1,3,5-triazin-2-yl) ureido sulfonyl) benzoate)
- sulfometuron (2-[3-(4,6-dimethyl-1,3-pyrimidin-2-yl)-ureidosulfonyl]-benzoate)

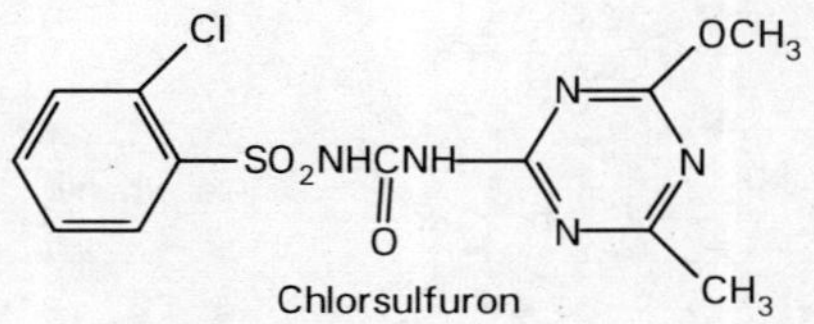

Chlorsulfuron

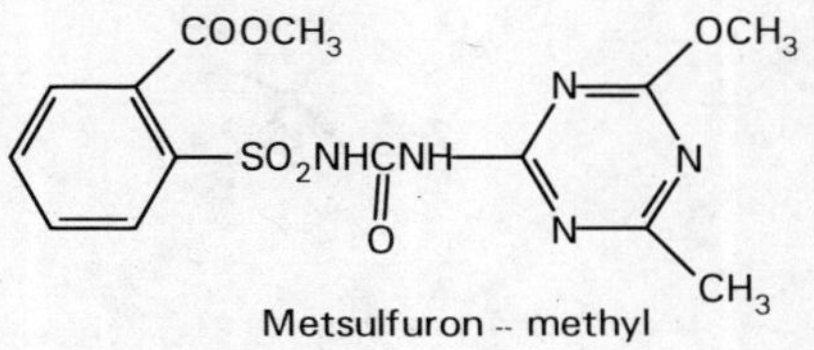

Metsulfuron -- methyl

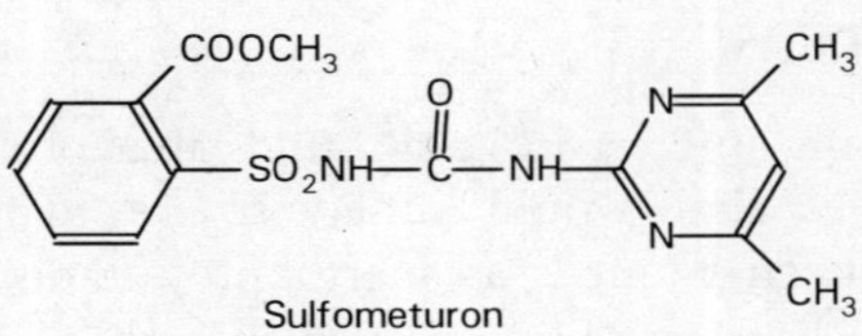

Sulfometuron

The high activity of the sulfonyl urea herbicides suggests that they act by a very specific biochemical mode of action. The herbicidal action of sulfonyl urea arises from inhibition of cell division, because this was demonstrated to be most sensitive process. Genetic and biochemical studies indicated that sulfonylureas inhibit the action of enzymes acetolactate synthase (ALS) which catalyses the first stage in the biosynthesis of the essential aminoacids valine and isoleucine. The primary biochemical target of the sulfonyl urea herbicides is therefore concluded to be enzyme ALS. The proposed biochemical mode of action is supported by the studies that showed that the addition of amino acids valine and isoleucine completely reversed the herbicidal action of sulfonyl ureas towards sensitive plants.

## IMDAZOLINONES

The imidazolinones represent a new class of herbicide developed by the American Cyanamid Company in 1983 for control of mono and dicotyledonous weeds. The imidazolinones have a primary role in weed control in many crops. Imazaquin is widely used to control many broad leaved weeds in soybeans. Imazethapyr is also highly effective herbicide for difficult control grass and broad leaved weeds in soybean as well as other legumes including beans, alfalfa and peanuts. Imazamethabenz-methyl is used to kill lusters and deciduous trees.

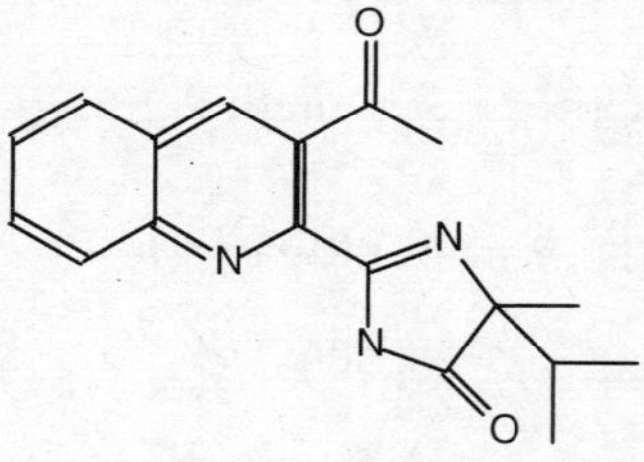

Imazaquin

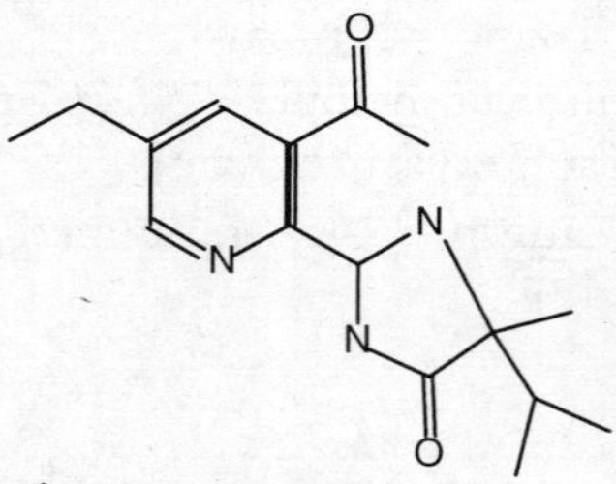

Imazethapyr

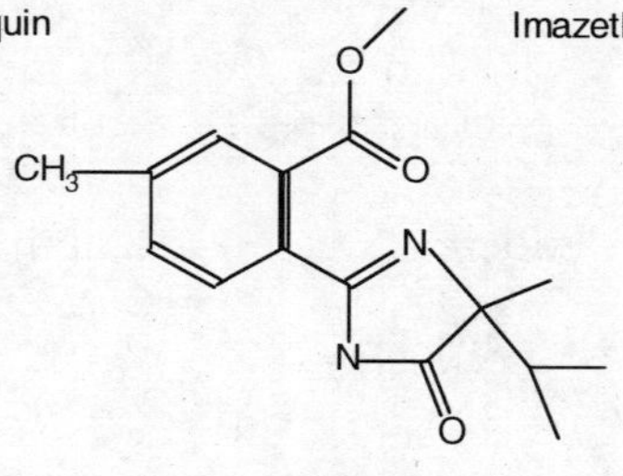

Imazamethabenz-methyl

## HERBICIDE EXTENDERS

There is an increasing interest in compounds that protect crops from the action of herbicides that would otherwise be toxic to them. These compounds have been termed 'safeners', protectants or with semantic accuracy 'antidotes'. They are used to treat the crops, before or alongwith the herbicide application. Compounds like diallyl dichloroacetamide, used in combination with EPTC, and cyometrinile, (dichlonaid). applied to sorghum seed to protect it from chloroacetamide herbicide injury. Napthalic anhydride and phathalic anhydride protected crops like sorghum and rice from injury by chlorosulfuron are already instituted and others are at various stages of development.

The introduction of crop safeners for herbicide have opened a new era in chemical weed control. They permit chemicals to be safely applied in higher doses and may allow the use of more potent herbicides, so giving a wider spectrum of weed control. To date, there has been little success in discovery of safener to protect crops against damage caused by photosynthesis inhibiting herbicides, paraquat or glyphosate. The majority of safeners are effective against soil herbicides especially thiocarbamates and chloroacetanilides.

The precise knowledge of mode of action of safeners remains incomlplete, however the dichloroacetanilides owe their specific safening action towards thiocarbamates and dichloroacetamides to their structural and chemical similarity to these herbicides. The mode of action of safeners with sulphonylurea herbicides in wheat and maize also appears to be associated with enhanced metabolism to herbicidal active products. The majority of safeners appear to act by increasing the rate of metabolism of herbicide, however, other possible mechanisms of action include interference with the uptake and translocation of the herbicide or effecting sensitivity of the target site.

## BIBLIOGRAPHY


1. Matolcsy, Gy, Nadasy M, Andriska V Pesticide chemistry, ELSEVIER, 1988
2. Ashton, F.M. Crafts, A.S. (1973) mode of action of Herbicides Wieley-Interscience New York, 1973.
3. Thompson W.T. Agricultural Chemicals Books. II Herbicides. Thompson Publ, 1975.
4. Hearney P.C. and Kayfman, Degradation of Herbicides, New York, Becher 1969.
5. Metcalf R.L. Adv. in Pest Control Research, Vol III, Interscience, New York, 1957.

# CHAPTER 12

# Plant Growth Regulators

Plant growth regulators are chemicals used for causing acceleration or retardation of the rate of growth or rate of maturation of the plants and are legally considered as pesticides. Hormones are natural substances produced by the plants that control growth, initiate flowering, cause blossoms to fall, set fruit, cause fruit and leaves to fall, control initiation and termination of dormancy and stimulate root development. The plant growth regulators are classfied into following groups:

1. auxins
2. gibberelins
3. cytokinins,
4. inhibitors and retardants
5. ethylene generators.

## AUXINS

Auxins are compounds that include elongation in shoot cells. The only naturally occurring auxin is indole acetic acid which probably occurs in all plant tissues. Important examples of auxins are:

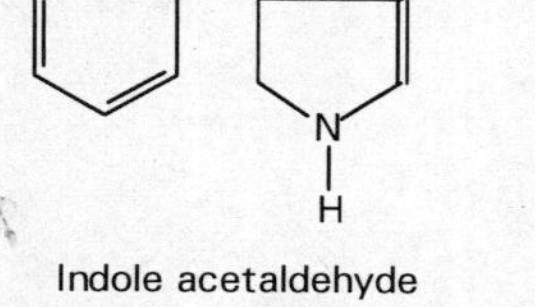


Indole acetic acid

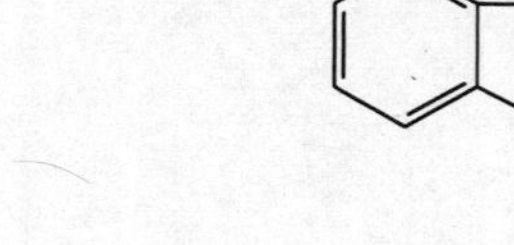


2,4-Dichlorophenoxy acetic acid

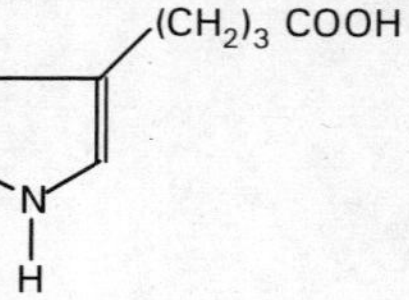


1-Naphthyl acetic acid

Indole acetaldehyde




3- indole butyric acid

## SYNTHETIC GROWTH INHIBITORS

### Maleic Hydrazide

It was the first synthetic plant growth regulator and it is used as a stunting agent to reduce rate of growth of grapes and some weeds on ranches to support root and sucker growth on tomatoes, tobacco and certain trees and to prevent sprouting of potatoes and onions in storage. Maleic hydrazide is prepared by maleic anhydride and hydrazine

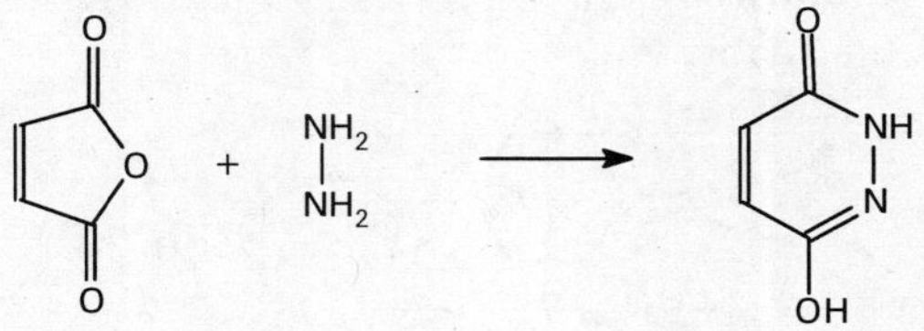

## Chlormequat

Chlormequat reduces stem length and lodging and produces compact plants with reduced internodes. It is used commercially in fruit trees and for reducing the height of cereals. It is prepared by treating ethylene chloride and trimethyl amine

$$ClCH_2CH_2Cl + N(CH_3)_3 \longrightarrow [ClCH_2CH_2N^+(CH_3)_3]Cl^-$$

## Daminozide

Daminozide is used to control the growth of fruit trees and shape and height of ornamentals. Daminozide is prepared by reaction of succinic anhydride and N,N-dimethyl hydrazine.

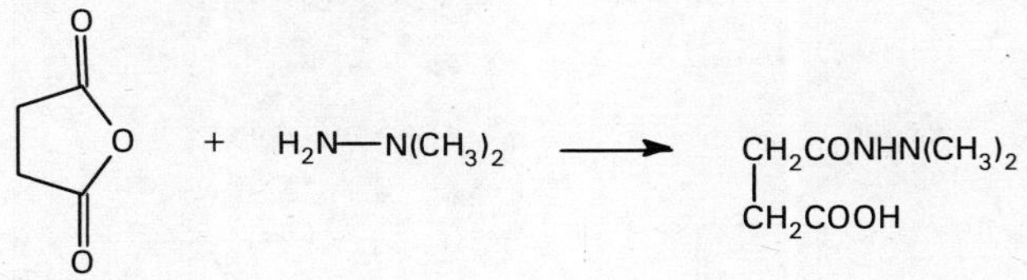

## Glyphosine

Glyphosine is used to increase the yield of sugar and hasten ripening

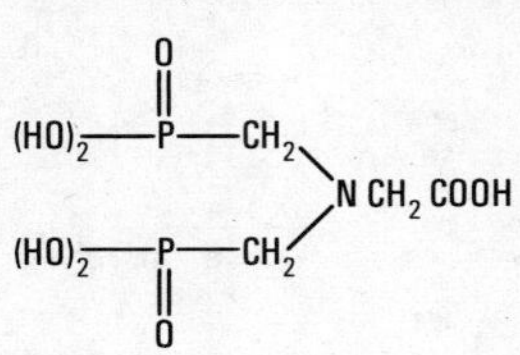

# GIBBERELLINS

Gibberellins are compounds that stimulate cell division or cell elongation and have a gibbane skeleton. Gibberellic acid was isolated from fungus *Gibberella fujikuroi*. Gibberellins were found to be natural constituents in all plants. More than 33 gibberellins have been isolated and are identified as GA1, GA2 etc. The gibberellic acid is the commonly used as GA3. The use of gibberellins are:

1. increase stalk length
2. yields of celery
3. to break down dormancy of potatoes seed
4. to delay fruit maturity
5. to improve fruit quality

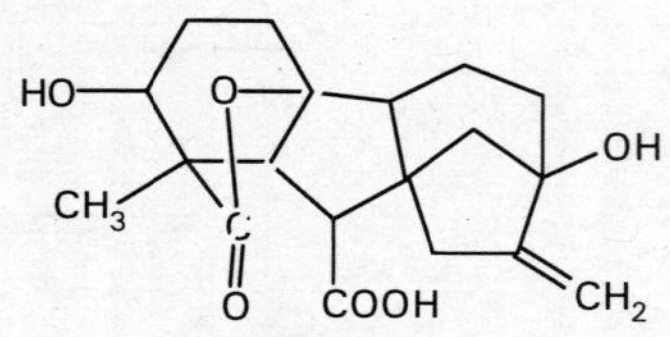

Gibberelic acid

## CYTOKININS

Cytokinin are compounds that induce cell division in plant and are used for prolonging the storage life of green vegetables, cut flowers, and mushrooms.

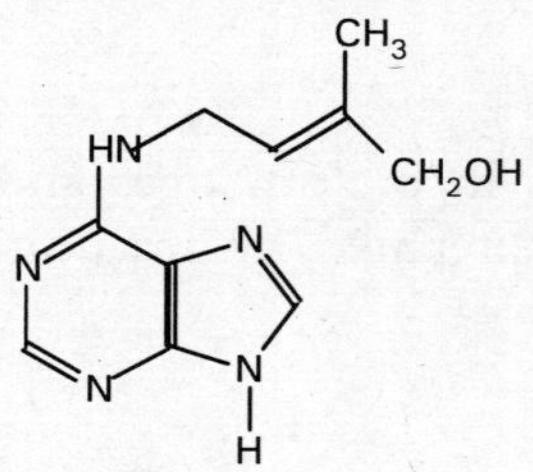

Zeatin

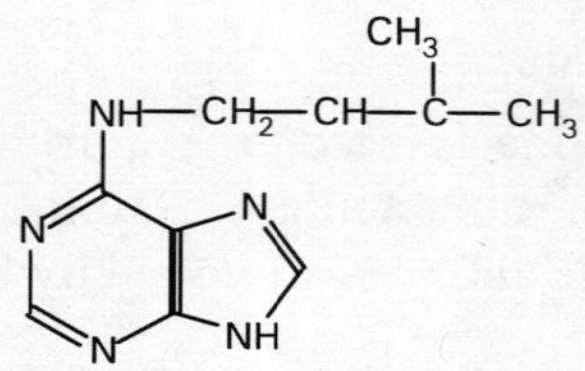

6 - isopentyl amino purine

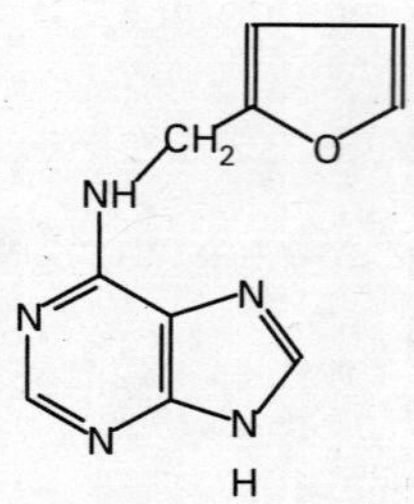

6- Furfuryl amino purine
Kinetin

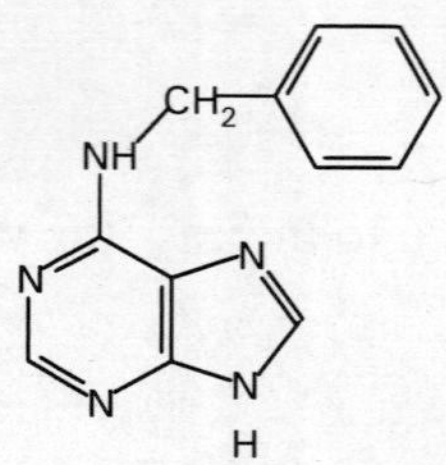

6 - benzyl amino purine

## INHIBITORS AND RETARDANTS

These are group of substances that inhibit a physiological process in plants. The use of inhibitors are:

1. They prevent sprouting of stored onions, potatoes and other root crops
2. They induce shortening of stems
3. They prevent lodging of grain and

4. They permit controlled growth of flower crops and ornaments.

Natural growth inhibitors exist in cotton, sycamore and yellow lupin. The cotton growth inhibitor abscisic acid also occurs in other plants.

Abscisic acid is a natural growth regulator one used to induce dormancy in cotton leaves, plum twigs and reduces germination in rose tips.

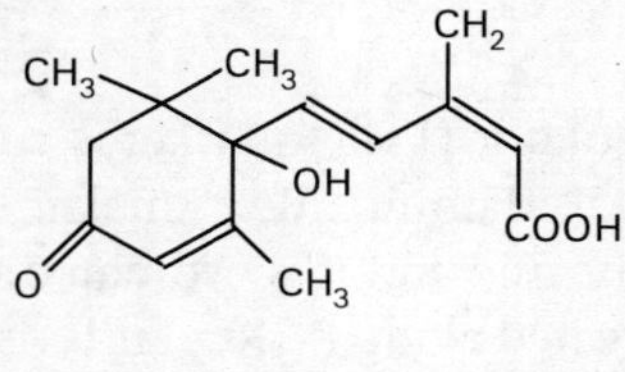

Abscisic acid

## ETHYLENE GENERATORS

Ethylene can inhibit growth, accelerate leaf and flower drop (abscission), the ripening of fruit and flowering depending upon the stage on which it is applied. Recently, chemicals have been developed which when sprayed onto plants release ethylene and hence control different phases of plants development.

Bisbenzoloxy 2-chloroethyl-methylsilane produces ethylene and is used as chemical abscission agent for apples, peaches and oranges and other fruits.

Ethepon (2-chloroethyl phosphonic acid) is prepared from phosphorous trichloride and ethylene oxide. Ethepon generates ethylene in solutions with pH > 4. It is sprayed on fruit trees to promote ripening, in rubber plantation to stimulate latex production and to improve stem strength in cereals.

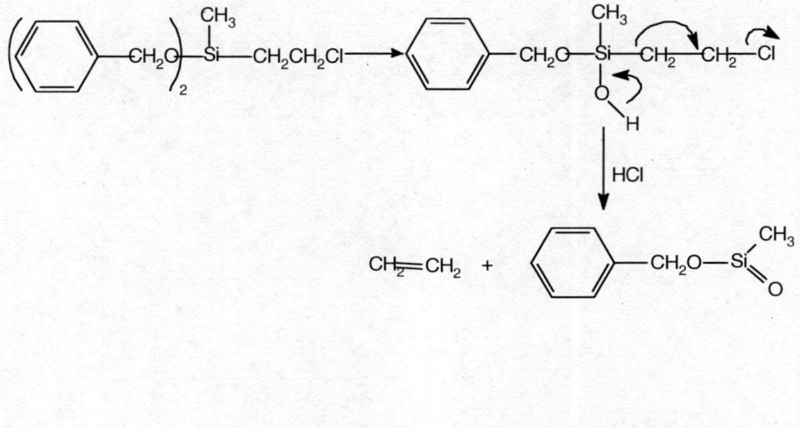

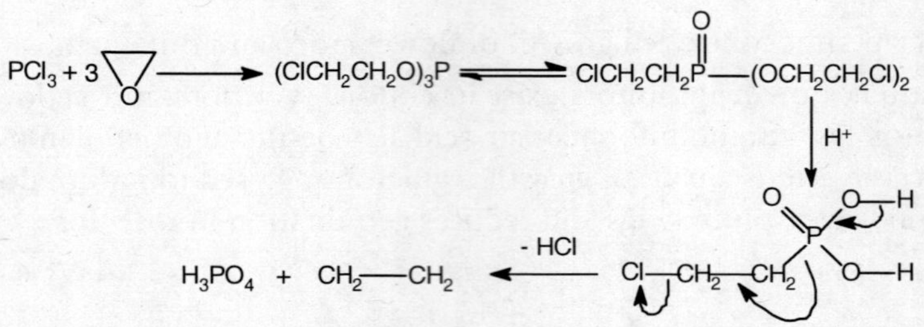

The plant growth regulator (PGR) market is currently dominated by six chemicals chlormequat, daminozide, maleic hydrazide, ethephon, gibberellic acid and glyphosine and it is estimated that one million hectares of crops are annually treated with PGRs. Little is known about mode of action of many PGRs and a better understanding of the mechanism of the structure activity relationships should permit introduction of more effective compounds.

## BIBLIOGRAPHY

1. Carr D.J. Plant Growth Substances, Springer Verlog 1970.
2. Wittwer S.H. Growth Regulants in Agriculture, outlooks on Agriculture, 1971.
3. Fleoning W.J. and Howden M.E.H. The chemistry of naturally occuring plant growth regulators Rev. Pure Appl. Chem 1972.

# CHAPTER 13

# Biological Control

The term 'biological control' has been defined by the National Academy of Sciences (NAS 1987) Research briefing Panel as the use of natural or modified organisms, genes or gene products to reduce the effects of undesirable organisms (pests) and to favour desirable organisms such as crops, trees, animals and beneficial insects and microorganisms. Only fungi, viruses, and bacteria are treated as insect pathogens. These pathogens all have one common factor that they cause diseases in other organisms when isolated from one diseased organisms, they are able to cause the same disease in other organisms.

Main advantages of biological control over chemical control are

1. The pest problems are neither intensified nor new ones created.
2. The beneficial organisms are already available i.e no manufacturing process is necessary.
3. beneficial organisms can increase in number and spread.
4. the pest will be unable or slow to develop resistance to such control.
5. the control is self perpetuating
6. leave no toxic residues
7. are compatible with toxic chemicals and can even sometimes be combined with them in a spray.
8. Have a low dosage requirement and are therefore, cheap and versatile.

The main disadvantages of biological control are:

1. Control is slow
2. It is difficult and expensive to develop
3. It requires expert supervision
4. Although not toxic to the consumer, the rotting cadavers of the diseased pests are an unacceptable additives to the producer

## Fungi

Fungi usually infect their hosts by direct invasion and are therefore able to attack the pests without being first ingested. A problem with fungi is the need for a fairly high humidity and temperature. If these two factors vary far from the optimum requirement of fungus, the action is slow and desired results are not obtained. This is the reason why several of the present fungal based commercial pesticides are targetted to glasshouse pests only. Some of the examples are that *Blacuveria bassiana* is already in extensive use for control of colorado potato beetle. The usefulness of some entomopathogenic fungi like *Verticillium lacani* for the control of soft green scale of coffee and citrus is also known.

## Viruses

Viruses are semicroscopic obligate, intercellular, pathogenic entities. Most of the viruses are infact not only insects, but also vertebrates or even plants. Multiplication of insect viruses is host specific. Insect viruses, the virions of which are embedded in polyhedral inclusion bodies, are called polyhyderal viruses. Extensive studies have been made with Nuclear Polyhedrosis Viruses (NPV) of *Heliothis armigera,* which is a polyphagous pest of many crops. Three to four sprayings of Polyhedral Occlusion bodies (POB) at 7-10 days interval at early stages effectively check the pest on chickpea, pigeonpea, cotton, sunflower and tomato. The various biological agents are given in the table 1.

**TABLE 14:** Various biological agents and their targeted site.

| Biocontrol agent | Target pests/diseased controlled |
|---|---|
| Egg parasitoid, *Trichogramma* | a) Shoot borer and internode borer of sugarcane<br>b) Heliothis on groundnut and cotton |
| Green lacewing fly *Chrysopa* | Eggs and small larvae of Helothis on cotton and groundnut |
| Antagonistic fungus *Trichoderma* | Root rot disease of groundnut, bengal gram and greengram |
| Exotic weevils, Neochitina eichhorniae and N. brunchi | Weed-water hyacinth |
| Beetle Zycogramma bicolorata | Weed-parathanium |

## Microbial Pesticides

This is biological control using insect diseases, a sort of germ warfare against insects. Only fungi and viruses are treated here as insect pathogens. These pathogens all have one factor in common; they cause disease in other foreign organisms. When isolated from one diseased organism, they are able to cause the same disease in other organims. The NPV was also reported to give excellent control of *Spodoptera litura* on cotton, tobacco, black gram, groundnut and cauliflower. Out of the various granulosis viruses (Gv), the Gv of sugarcane shoot borer, *Chiloinfuscatellus* is found effective in controling the shoot borer damage and increasing the yield. The significance of viruses in microbiological control has dropped considerably since the development of effective products based on *Bacillus thuringensis*.

## Bacteria

Bacteria are ubiutious in nature and have a versatile metabolism, which allows them to live in every part of the biosphere. The majority of bacterial strains which have been closely examined as insect control agents and have been used for commercial purposes, belong to the genus *Bacillus* e.g. *Bacillus thuringiensis, Bacillus popillae, Bacillus sphaericus. Bacillus thuringrensis* (Bt) refers to a group of bacterial isolates which was first isolated in 1915 from diseased flour moth larvae and later several species were obtained from different species of infected caterpillars. It is marketed under the trade name Thuricide, Dipal, Agritol and Dandro bacillin etc.

*Bacillus thuringensis* was initially believed to kill larvae by septicaemia i.e. by multiplication of the vegetable cells in the gut and in the body cavity. It is now well established that delta-endotoxin alone is responsible for the death of most susceptible lepidopterous larve. Septicaemia may play a role in addition to poisoning by the delta-endotoxin only in minority of species. After ingestion by the susceptible larvae, the crystal protein broken by midgut juice proteases, under high pH > 9 conditions, into smaller toxic peptic molecules, the delta-endotoxin. The latter causes mouth and gut paralysis within 1/2 hour of eating a large dose, thus preventing further feeding within hours, the epithelium of the midgut is destroyed and gut contents invade the body cavity, rapidly causing death. However this protein of *Bacillus thuringiensis* is non toxic to verteberates and development of commercial products based on *B. thuringiensis* has given a significant impulse to microbiolgical pest control). With certain strains of *B. thuringiensis*, toxin known as B-ecotoxin can be isolated from their cultures B-ecotoxic is thermally stable and possesses much wider spectrum of insecticidal activity, affecting larval stages of many species of insects and also showing toxicity to range of mammals.

## BIBLIOGRAPHY

1. Hagen K.S. and Franz J.M. A History of Biological Control in History of Entomology pp 433-476 Ann. Rev. Inc. Pub Alto 14 SA 1993.
2. Sweetman H.L., The principles of biological control-interrelation of hosts and pests and utilization in regulation of animal and plant population pp 1-560 wm. C. Brown Dubugue, Iowo 1958.
3. Iqnoffo C.M., Biever K.D. and Hostetler living Insecticides Chem. Technol, 1975.

# CHAPTER 14

# Repellents, Attractants, Pheromones, Synergists, Synthetic Inhibitors and Proinsecticides

## REPELLENTS

Repellents are chemicals which cause insects to make oriented movements away from its source. The first chemical repellent was discovered soon after man became acquainted with fire. When he found that fire was a fairly efficient method of repelling insects. Plant extracts of citronella, which is extracted from *Andropogm nardus* (L.), which contains gerniol, cintronellol, cintronellal, borneol, and terpenes are considered to be the principal mosquito repellents. Now more effective, synthetic repellents are in use.

### Dimethyl Phthalate

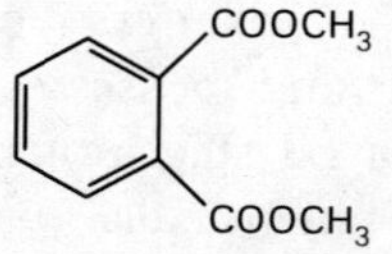


Dimethylphthalate is used as effective mosquito repellent. The disadvantage of this repellent is that it causes irritation to the eyes and mucous membranes. Its $LD_{50}$ value is 8200 mg/kg.

### N,N-Diethyl-m-toluamide (Deet)

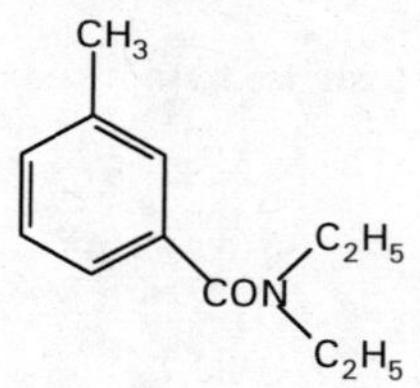


Deet is used as repellent for various insects particularly to mosquitoes. Its $LD_{50}$ value is 200 mg/kg.

## Benzyl Benzoate

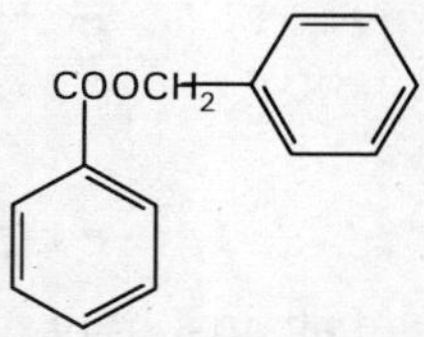

Benzyl benzoate is used as repellent for clothing treatment against chiggers and mites. Its $LD_{50}$ value is 7900 mg/kg.

## Butoxypolyproplene Glycol

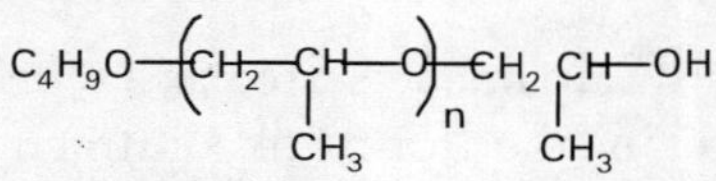

Butoxypolyproplene glycol is used as repellent for flies against cattle.

## 1,5a, 6,9,9a,9b-hexahydro-4a (4H) dibenzofuran-carboxaldehyde (MGK, Repellent 11)

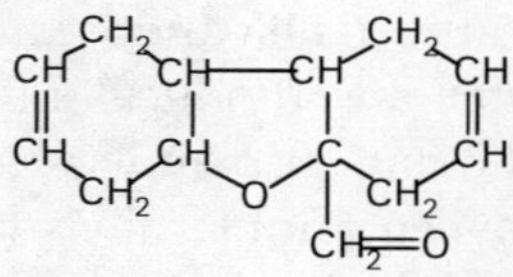

MGK repellent used as repellent for cockroaches and for biting flies on cattle. Its $LD_{50}$ value is 2500 mg/kg.

## 2-Ethyl-1,3-hexanediol (Rutgers 612)

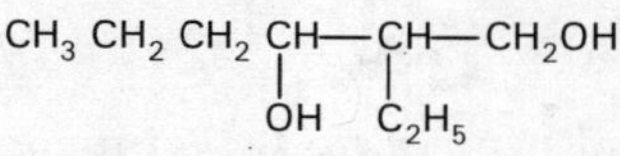

Ruters 612 is used as repellent against flies and mosquitoes.

## Butyl 3,4-dihydro-2,2-dimethyl-4-oxo-2H-pyran-6-carboxylate (Indalone)

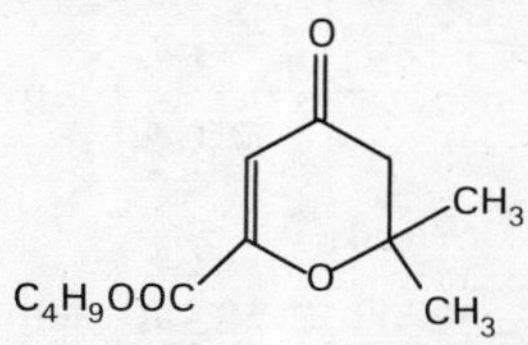

Indalone is used as repellent against mosquitoes and bitting insects. Its $LD_{50}$ value is 7800 mg/kg.

## Di-n-propyl 2,5-pyridine dicarboxylate (MGK 326)

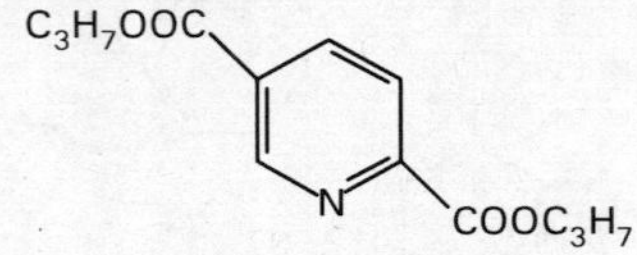

MGK 326 is used as repellent against housefly.

## n-Propyl-N-N-diethylsuccinamate

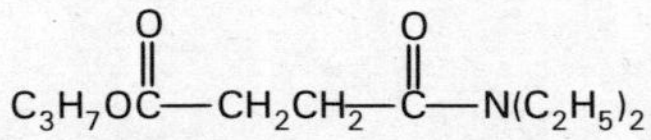

n-propyl-N-N-diethylsuccinamate is used as a repellent for mosquitoes.

Repellents are used in the form of ointments, pastes, solutions, emulsions and acrosols.

## ATRACTANTS AND PHEROMONES

Attractants are chemical substances whose vapours attract insects. They are mostly divided into food attractants and sexual attractants. Important examples of food attractants are products of decomposition of food products, proteins, enzymatic and also molasses. Some of the important synthetic attractants are methyl eugenol (1-allyl 3,4-dimethoxybenzene),

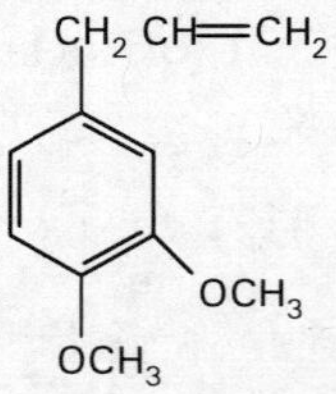

attracting estern fly and trimedlure (1,1-dimethylethyl, 4 (or 5) chloro-2- methyl (cyclohexane) carboxylate, an attarctant for melon fly.

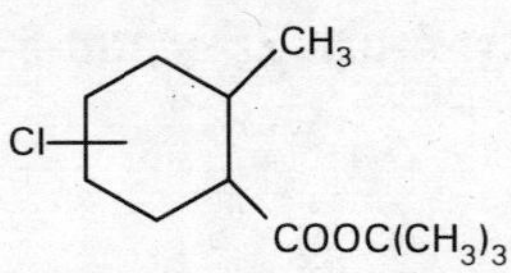

Trimedlure

There are few examples of synthetic attractants that are effective for using over fruit flies. Butyl sorbate is an effective for base for European cockroches and methyl linolenate for bark beetle.

## PHEROMONES (SEX ATTRACTANTS)

Pheromones are defined as those chemicals that are secreted into the external environment by an animal and that elicit a specific reaction in a receiving individual of the same species.

Sex pheromones are those pheromones which are released by one sex only and trigger behaviour patters in other sex that facilitate mating. Sex pheromones are produced by males of certain insect species. In most cases they serve as short-range mating stimulants after the individuals have come into close proximity.

**Chemistry** The identified sex pheromones are:

***Silkworm:*** trans-10-cis-12-hexadecadienol,

$$CH_3(CH_2)_2CH = CHCHCH{=}(CH_2)_8CH_2OH$$

***Gypsy moth, Lymantria obfuscata:*** 10-acetoxy-cis-7-hexadecenol

$$CH_3(C_2H_5)_2\underset{\displaystyle OCOCH_3}{\underset{|}{C}H}-CH_2CH{=}CH(CH_2)_5CH_2OH$$

***Pink boll worm, Pectinophera gossypiella:*** 10-propyl-trans-5,9-tridicadienyl acetate

$$(CH_3(CH_2)_2)_2C{=}CH(CH_2)CH{=}CHO(CH_2)_3(CH_2)_6O\overset{\displaystyle O}{\overset{\|}{C}}-CH_3$$

Cabbage looper, *Trichloplusia* ni (Huubner) - Cis-7- dodecenyl acetate

$$(CH_3(CH_2)_3CH{=}CH(CH_2)_5CH_2O\underset{\displaystyle O}{\underset{\|}{C}}-CH_3$$

Honey bee Queen, *Aphis mellifera* L. 9-keto-trans-2-decenoic acid

$$CH_3-\underset{\displaystyle O}{\underset{\|}{C}}-(CH_2)_5\,CH{=}CH-COOH$$

Aggregation pheromones, are those pheromones which may be released only by one sex but cause approach responses by individual of both sexes of the species.

## Uses of Pheromones

1. Pheromones can be used to attract insects to a site where they are then destroyed by treatment with insecticides
2. The use of pheromones enables pests to be controlled specifically. The minute application quantities avoid contamination of food and fodder as well as environment.
3. Pheromones have great potential usefulness in surveys to determine the presence of abundance of species in question, so that other measures can be excercised.

# SYNERGISTS

Synergist may be defined as any chemical which itself is not toxic to insects at dosages used but when combined with an insecticide greatly enhance the toxicity of the insecticide. Synergism differs from poteniation which occurs when two toxic materials applied together elicit a response greater than that expected from the sum of the individual toxicants.

Advantages of using synergists

1. It reduces the amount of pesticide required for insect kill, much smaller application of pesticides would be required for field application.
2. It reduces the environmental pollution.
3. It broadens the spectrum of the insecticide. Of the several different groups of chemicals able to synergise insecticides, the most widely studied are methylenedioxyphenyl derivatives. Important examples are piperonyl butoxide, safrole, sulfoxide, tropitol and thanite.

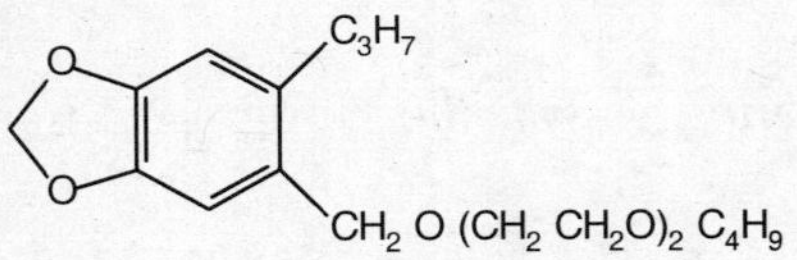


Piperonyl butoxide

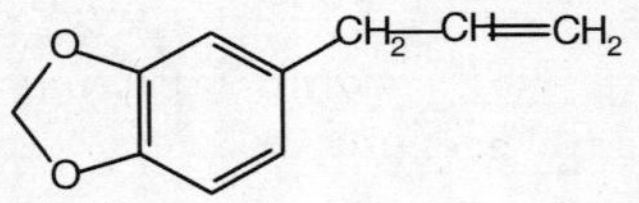


Safrole

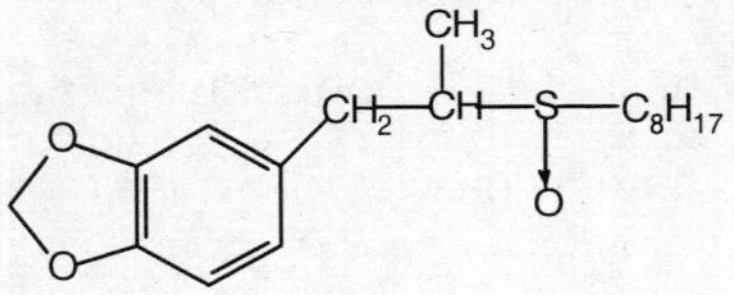


Sulfoxide

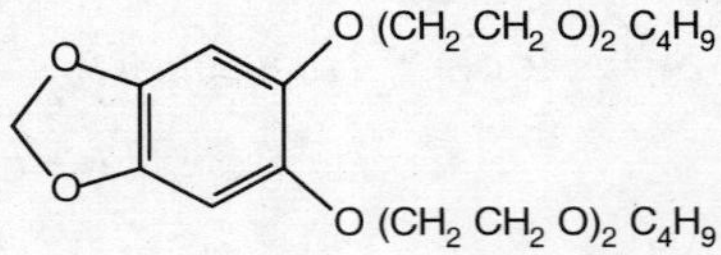


Tropitol

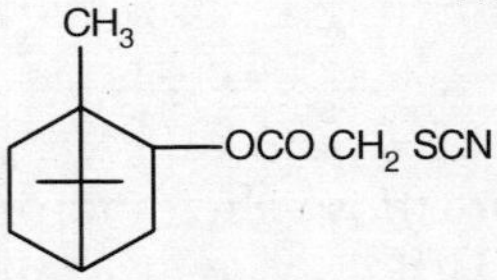


Thanite

## Autosynergism

Autosynergist is a compound which synergises its own toxicity. This would involve forging a synergophore and toxophore into the same molecule in such a way that the former does not interfere with the action of the latter. Important example of autosynergism in 3,4-methylenedioxy phenyl N-methyl carbamate

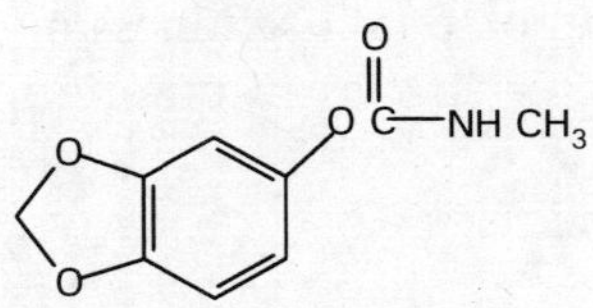


## SYNTHETIC INHIBITORS

Inhibitors are organic or inorganic substances of a varied chemical nature, and also the products of metabolism of a cell, that cause the partial or complete suppression of the activity of enzymes or metabolism in a living organism.

Chitin, essentially a polymer of N-acetyl-glucosamine, is a structural feature of the insect cuticule and fungal cell walls, hence its biosynthesis is an attractive target for selective fungicides. Important example of chitin inhibitor is diflubenzuron. Diflubenzuron is highly active against mosquitoes, livestock fly pests, cabbage white butterfly caterpillars, moths and other larval fly pests at low dosages. The biochemical basis of action is considered to be disruption of the moulting process by inhibition of the normal insect cuticle formation. Diflubenzuron is prepared by the reaction of 2,6-dichlorobenzamide with p-chlorophenylisocyanate.

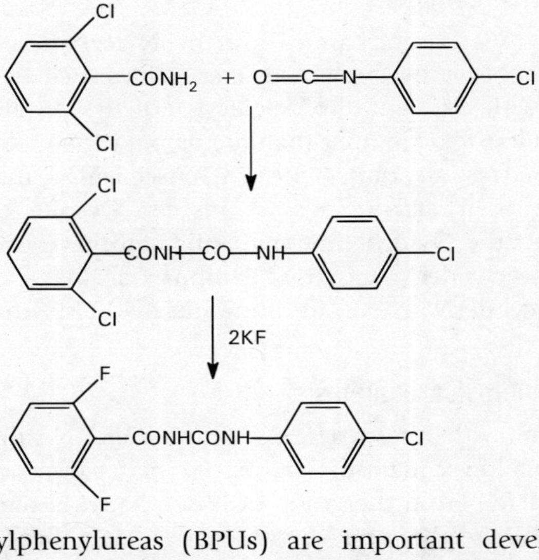


The benzoylphenylureas (BPUs) are important developing class of chitin inhibitors. Flufenoxuron and flucycloxuron are important insecticides belonging to the form and are active against mites as well as insects. The benzoylphenylureas, all act by the interference with chitin synthesis in immature insects and are highly selective insecticide with low toxicity to target organisms.

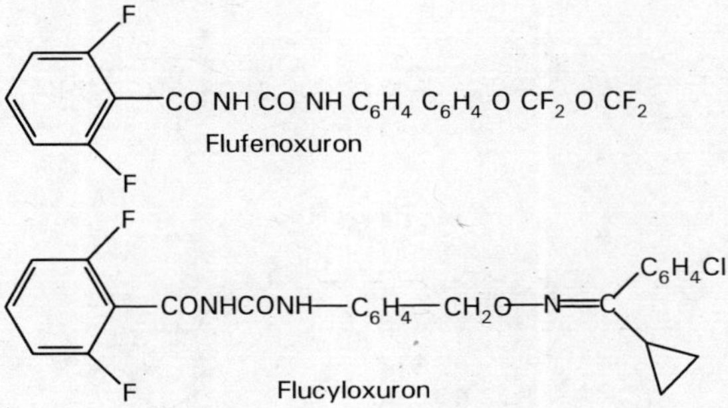


Flufenoxuron

Flucyloxuron

## PROINSECTICIDES

In 1958, Albert introduced the term prodrug, a substance which has to be modified to become a true drug. By analogy, substances which have to be transformed in order to show or to enhance their insecticidal activity are known as pro-insecticides. Such compounds can be activated in various ways. The process can be non catalytic such as hydrolysis, thiolysis, elimination, photolysis, cyclization, rearrangement etc. or by enzymatic catalysis, catalysed by oxidases, carboxylases, amidases, peptidases etc.

## PROCARBAMATES

### N-acyl-N-methylcarbamates

The oldest group of procarbamtes are the N-acyl-N- methylcarbamates. Derivatisation of N methylcarbamates insecticides such as carbaryl, landrin and propoxur with different aliphatic acyl moieties produced compounds that were much less toxic to mice than the parent carbamates. The N-acyl-N-methyl carbamates are quite effective insecticides than their parent compounds in their activity against insects. Despite their insecticidal activity, none of the N-acyl-N-methyl carbamate have been developed as commercial insecticides. Recently, Union Carbide examined a new compound N-acyl derivative of carbofuran in field tests under the code No. UC54229.

### N-sulphenyl-N-methylcarbamates

In comparision with their parent compounds, N-sulphenyl-N-methyl carbamates show lower mammalian toxicity, better insecticidal activity and lower phytotoxicity. From the group of N-sulphenyl-N methyl carbamates, carbosulfan, thiodicarb, and furathiocarb have been developed to commercial insecticides.

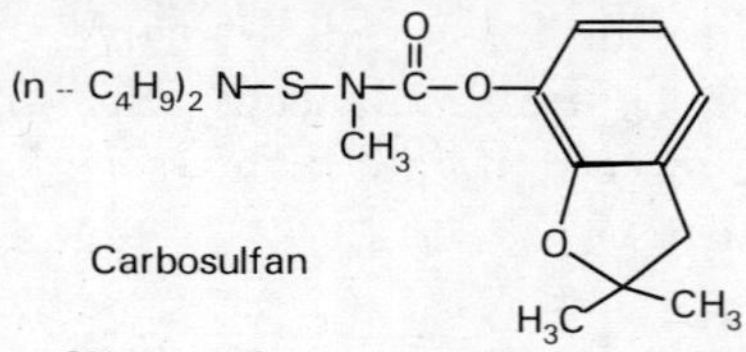

Carbosulfan

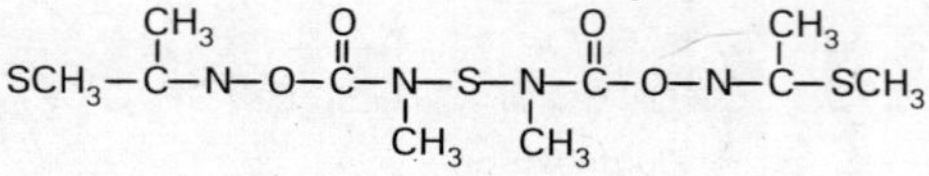

Thiodicarb

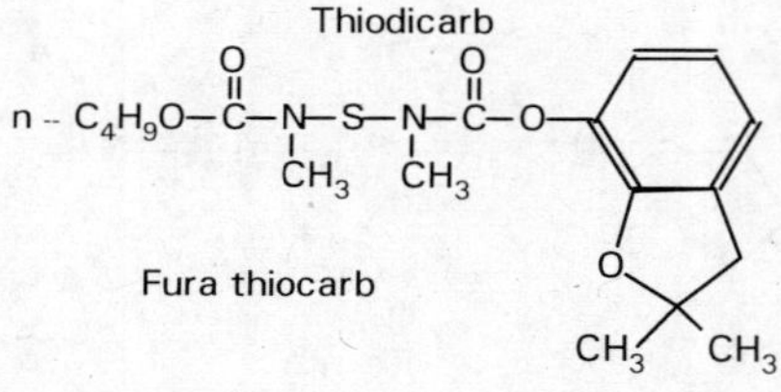

Fura thiocarb

## Propyrethroids

On the vinyl side chain of the acid moiety brominated derivatives of permethrin, cypermethrin and deltamethrin, show insecticidal activity which on molecular basis is comparable to that of the corresponding dihalovinyl analogues. This activity is hardly compatible with the structure because the presence of the vinyl group in the acid moiety determined to a great extent the biological activity of such esters.

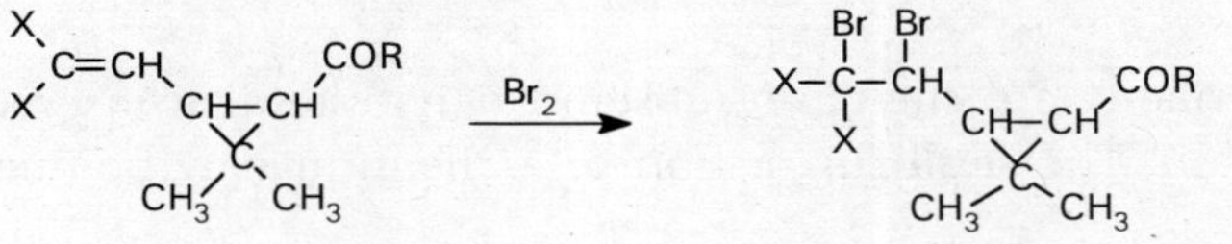

X = Cl, Br R = OH, Cl or ester group

It is known that the organophosphorus insecticides is transformed in the presence of metals, SH groups or other reducing groups to dichlorovos, which is the actual toxicant. It was expected that similar debromination which leads to the vinyl esters could be responsible for the high insecticidal activity of brominated pyrethroids. *In vitro* brominated cypermethrin and deltamethrin can be debrominated to cypermethrin and deltamethrin by cystein or $H_2S$.

## Proinsecticidal Organophosphates

In the pure state, phosphorothionate esters are poor inhibitors of cholinestrases. The insecticidal activity of the thiono compounds is the result of the oxidative desulphuration of the thionophosphoryl group *in vitro*. The conversion of thiono into oxo esters in animals is caused by the microsomal mixed function oxidase systems. In plants, per-oxidases and/or light may play a role in the thiono-oxo-transformation or oxidative desulphoration of the thioplosphoryl sulphur has been demonstrated such as parathion and malathion.

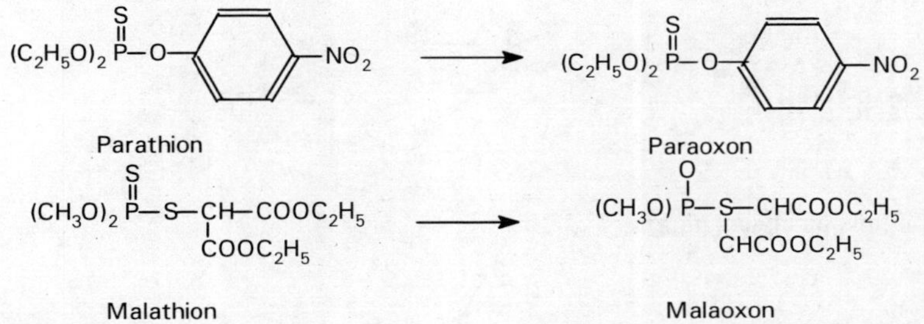

Parathion Paraoxon

Malathion Malaoxon

## BIBLIOGRAPHY


1. Büchel K.H. Inseckten - Repellents Springer Verlog 1970.
2. Painter R.R. Pest Control, Acoolemic Press New York, 1967.
3. Jackbson M. Insect Sex Pheromes Academic Press, New York 1972.
4. Ibraski M.T. EI. Insect harmones and Analogues 2, Agnew - Entomol 66, 1970.

# CHAPTER 15

# Chemosterilants

Chemosterilants are chemical substances that sterilize insects to prevent reproduction. The sterilizing action of a chemical may be due to the fact that

1. It damages the chromosomes and leads to the lethal mutations in the semen of the males or matures eggs of females.
2. It inactivates the sperum
3. Under the influence of chemosterilants, females lose the ability of laying eggs.
4. It causes perishing of the cells in the premeiotic stages or aspermia

Most chemosterillants relate to highly toxic compounds with a mutagenic, tetragenic, and carcinogenic action.

Chemosterilants are divided into three groups:

1. Alkylating agents
2. Antimetabolites
3. Miscellaneous.

**Alkylating agents** are compounds in which hydrogen atom of molecule of a substance is replaced by an alkyl group. They readily enter into the chemical alkylation reactions with various compounds of a cell, including proteins and nucleic acids. The most important compounds of this group are

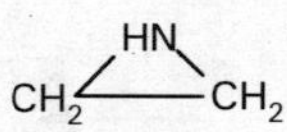


## ETHYLENE IMINE

### Tepa

(1,1',1"-Phosphinylidynetrisaziridine)

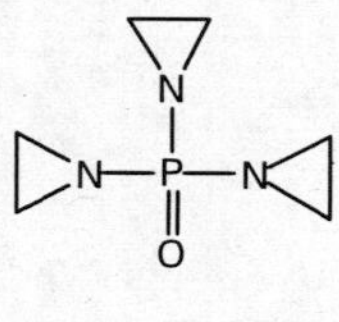


### Apholate

(2,2,4,4,6,6-Hexakis (1-aziridinyl)-2,2,4,4,6,6-hexahydro-1,3,5,2,4,6,

triazatriphosphorine)

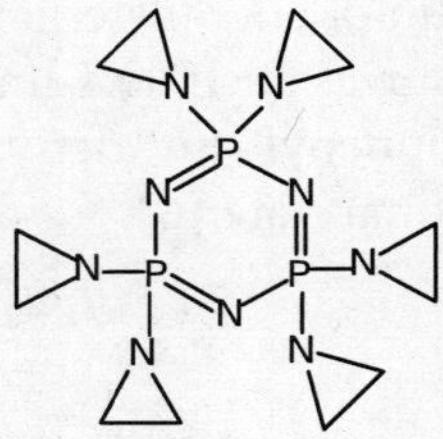

**Antimetabolites** are substances whose structure is very close to that of the natural metabolites of an organism. When they enter the organism they displace these metabolites in exchange reactions. Upon entering an insect organism, these substances violates the synthesis of nucleic acid (DNA and RNA) in the nuclei of the sexual cells. Among the anti-metabolites of folic acid, the best examples are methotrexate (N-p-2,4-diamino-6-pteridyl) methyl methylaminobenzoyl) glutamic acid), aminopterin (2,4 - diaminoptereldyl glutamic acid) and fluorouracil very close to the structure of folic acid.

Aminopterin

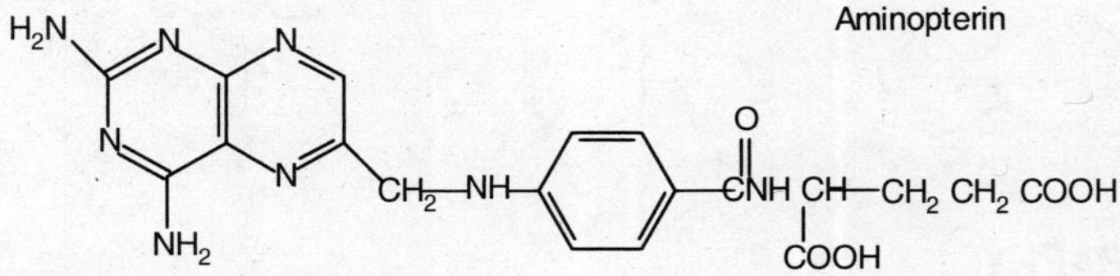

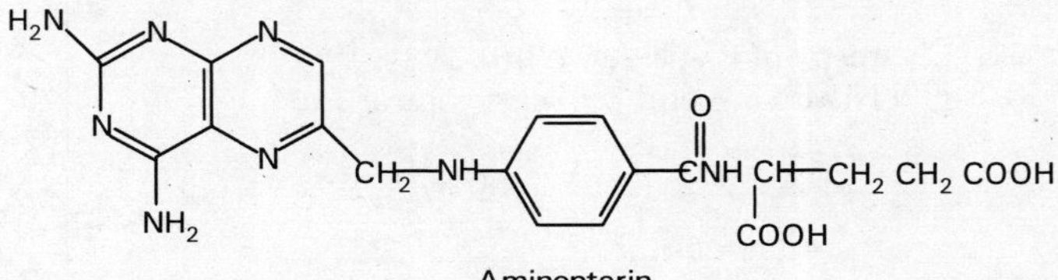

Aminopterin

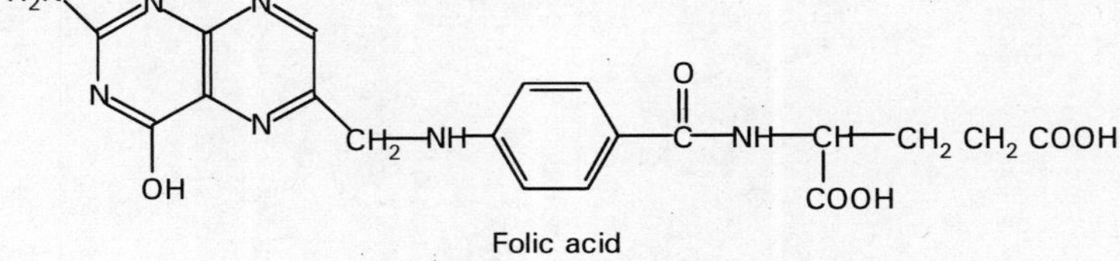

Folic acid

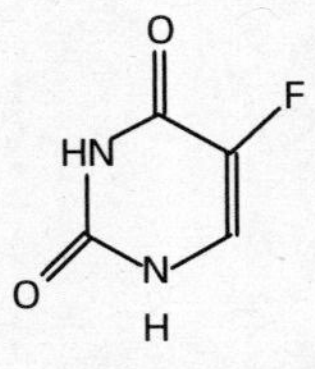

5-thuorouracil

## MISCELLANEOUS COMPOUNDS

The third miscellaneous group of chemosterilants are hempa (1) and hemel (2), urea and thiourea derivatives such as 4-imidazolin-2-one (3) and 1-(1-napthyl)-2-thiourea (4), triphenyl tin derivatives, such as triphenyltin hydroxide (5) and triphenyltinacetate (6).

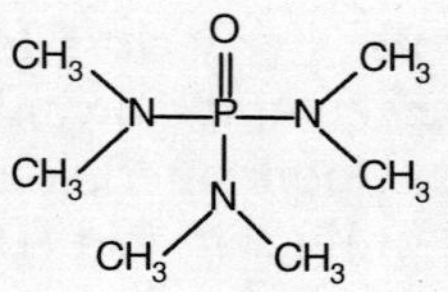


1

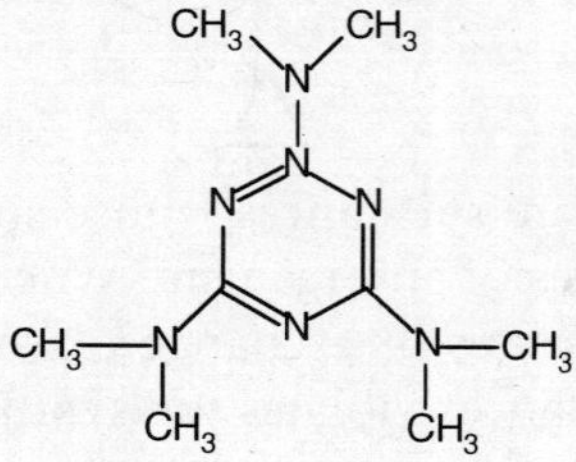


2

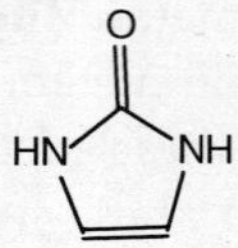


3

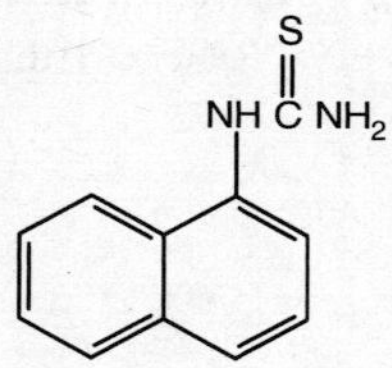


4

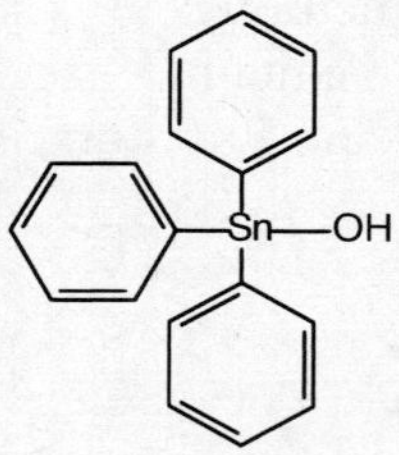


5

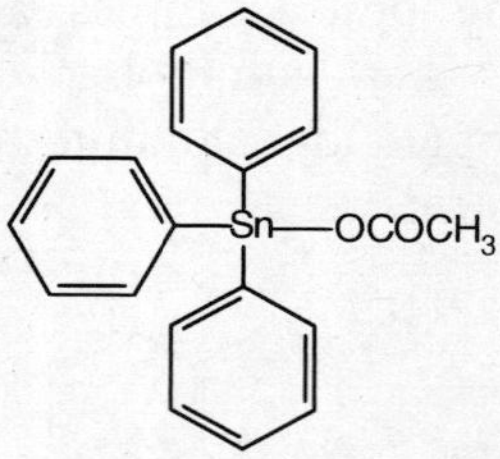


6

## BIBLIOGRAPHY

1. Büchel, K.H. Chemistry of Pesticides, Wiley, 1983.
2. Ascher, KL.S. and Nisrim S World Fer Penst control 1964.

# CHAPTER 16

# Pesticides and Environment

Pesticides are invaluable inputs for increasing agricultural production due to the fact that pests and diseases destroy upto one third of the crop during growth, harvest and storage. It is generally agreed that pesticides will be needed indefinately and their use will expand, inspite of the exploitation of alternate methods of pest control. However, the rapidly increasing usage of pesticides often with insufficient advice on research has brought in its wake many environmental problems inimical to the interest of man.

The use of pesticides leads to the residues on crop at harvest. Pesticide residues means any specified substance in food, agricultural commodities or animal feed, resulting from the use of pesticides. The term includes any derivative such as conversion products, metabolites, reaction products and impurities to be of toxicological significance. Pesticide residues may be divided into bound residues and conjugated residues. Bound residues may be defined as chemical species originating from pesticide usage that cannot be extracted by the methods commonly used in the analysis. Conjugated residues are products of secondary metabolism which involves the reaction of the pesticide or its metabolites with endogenous substrates such as sugars, gluconic acid, amino acids. The conjugated residues are generally extractable from the substrate with polar solvents.

## THE PERSISTENCE OF PESTICIDES IN PLANTS AND SOILS

It is very difficult to make generalizations as to the relative importance of different factors which influence the persistence of pesticides in plants and soils. However, the persistence of pesticides in plant and soil will depend on the following factors.

### Chemical Factors

1. **Structural stability**: The basic structure of the pesticides is fundamental in influencing their persistence in plants and soils. The chlorinated pesticides are highly persistent while the organophosphates and carbamates are not persistent in nature. Some pesticides can be partially broken down and then become extremely persistent as when aldrin is expoxidised to dieldrin and heptachlor to heptachlor epoxide, both these products being more stable than the parent compounds.
2. **Volatilization:** Pesticides range in volatility from extremely low to the

very high vapour pressures. It has been realized that there is a strong co-relation between the rates of loss of pesticides from plant surfaces and their vapour pressures.

3. **Solubility:** Solubility greatly influences the persistence of pesticides residue on plants. Soluble spray can be washed off from a crop within 24 hours, unless it is absorbed on to plant surface. There is also a good co-relation between the water solubility of pesticide and their persistence in soils. Soil is an excellent filter of fine particulates matter and there can be little carriage of undissolved pesticides in leachates. However, soluble pesticides pass down through the soil in large quantities to water drains downwards.
4. **Method of application:** The persistence of a particular pesticide depends greatly on the method by which it is applied. The gradual increase in efficiency of the equipment for applying pesticides has led to corresponding decrease in the dosage of actual toxicity per unit area.
5. **Formulations:** The ways in which pesticides are formulated considerably influence their persistence. Formulations in order of increasing persistence on plants and increasing size of particles are as water soluble liquids, emulsions, wettable powders, dusts, and granules. In soils, the rate of adsorption on to various soil fractions, particulatrly the organic mater, is inversely related to the particle size of the pesticides in the formulation. Hence, although water soluble formulations are not persistent in soils, because they can be easily leached away, emulsions tend to persist longer than the wettable powders or dusts, because they are adsorbed more readily on the soil fractions and so not appreciably degraded.

### Plant Factors

Plant species: The structural factors that influences the distribution, retention and uptake of pesticides have been listed as

1. Plant form (erect, spreading, prostrate)
2. leaf shape (broad, narrow, large, short linear)
3. leaf position and density (horizontal, upright, pendulous)
4. leaf surface and margins (hairy, waxy, sculptured).

Thus specific plant differences considerably affect the uptake and persistence of insecticide. The leaves of different species of plants differ greatly in wettability, and within a given species, immature leaves tend to be more difficult to wet than mature ones. The density of foliage is another important factor, some species having such a dense growth that it is virtually impossible for sprays to reach the inner most leaves.

## EASE OF PENETRATION

Pesticides remain on the surface of the plant organs disappear quite rapidly through various forms of weathering. For many pesticides, the case with which they are taken up into plants, depend upon wettability of the cuticle, but this changes with ageing and differs on different parts of the leaf and usually monocotyledenous plant can be wetted less easily than dicotyledenous ones. Roots absorb lipid soluble pesticides much less readily than water soluble ones, but leaf absorption differs, because absorption of pesticides by leaves is usually hindered by wax particles, hairs, thorns or protuberones, that produce air prockets which prevent the spread of pesticides.

## Metabolic Activity

1. **Rate of growth:** The rate at which a crop grows greatly influences the persistence of pesticide residues on the leaves and other parts. This is obvious because residues become spread over on much greater surface areas as the leaves and organs grow and become much more exposed to weathening. Moreover, if a plant doubles its weight, the pesticide residue concentration is halved, even if there is no increase in the rate of its breakdown.
2. **Excretion:** Pesticides can be excreted from various plant organs in guttation droplets from leaves or in root exudates. No doubt these phenomena are quite common but, it is doubtful whether very large pesticides are lost by excretion.

## Soil Factors

1. **Soil structure:** With a few exceptions, the smaller the particles a soil is composed of, the longer pesticides persist in it. Pesticides became readily absorbed on to soil particles thus making the pesticide persist much longer than it otherwise would, and include
   a) physical adsoption
   b) chemical adsorption
   c) hydrogen bonding
   d) Co-ordination (metal complexes)

   Soil structure also affects leachibility of pesticides because the pore size and pore size distribution greatly affect the movement of water through soil.
2. **Clay Content:** Clay is one of the most important soil constituents influencing the persistence of pesticides in soils. Such soils have a much larger internal reactive surface than other soils, thus providing a greater surface area of adsorption of pesticides. The factors that can influence the amount of pesticide sorbed by soil colloids are:
   a) the physiochemical configuration of the pesticide
   b) the dissociation constant of the pesticide
   c) the water solubility of the pesticide
   d) the size of the pesticide molecule

e) the soil hydrogen ion concentration
f) temperature
g) the electrical potential of clay surface
h) the moisture content of the soil
i) formulation

3. **Organic matter content:** The amount of organic matter in soils ranges from less than 1% to more than 50%. There is now much evidence to show that soil organic mater is extremely important in absorbing pesticides and many workers have reported strong co-relation between the persistence of pesticides in soil and the amount of organic matter in them. Most of the soil organic mater consists of humic compounds and have functional groups such as carboxyl, amino and phenolic hydroxyl groups, which provide sites for hydrogen bonding with pesticide molecules. Together, these characterstics provide considerable capacity for pesticide absorption and increased persistence.

## Metabolic Activity

1. **Soil microorganisms:** There seems little doubt that microbial activity is a major factor in the degradation of pesticides in soil. Many workers have reported that accelerated degradation of pesticides occurs when additional sources of energy are added to soils. It has been shown that populations of some species of microorganism increase after repeated aplications of pesticides to soils but it is not yet clear whether such increase occurs because the pesticides are a source of food for them or pesticides eliminates certain competing species.
2. **Free soil enzymes:** Enzymes are of biological origin but some can exist for considerable periods outside living organisms. They may be released from dying organisms, from microorganisms, from roots of plants or in excreta from soil animals and still remain active. It has been observed that soil enzymes are responsible for over all pesticide degradation.

## Environmental Factors

The influence of the climatic factors e.g., temperature, precipitation, radiation, and wind on pesticide breakdown is extremely complex, not only because, they effect most of the chemical, plant and soil factors, but they also interact with each other.

1. **Temperature:** All of the plant, soil and chemical processes of pesticide breakdown are affected by change in temperature. Higher temperature increases the rate of pesticides volatalisation from soil or plant surfaces and also their water solubility. The latter increases the rates at which the pesticide break down on the plants, penetrate the plant cuticle, or are absorbed by the roots and also rate of growth and translocation. All these factors accelerate the degradation of pesticides.
2. **Radiation:** Radiation increases the volatilization of most pesticides

from leaf surfaces. Even the most persistent pesticide such as DDT and HCH will degrade, to some extent, under the influence of radiation.

3. **Air movement:** The movement of air in the form of wind can effect the persistence of pesticides in plant and soil. Moreover, there may also be physical removal of deposits or sprays from surfaces of plants by wind and soil.

The widespread use of pesticides has resulted in problems caused by interaction with biological systems.

## HOW PESTICIDES POLLUTE THE ENVIRONMENT

### Exposure through Soil

Pesticides may reach the soil becsause of

1. Application of pesticide to soil for control of soil infesting insects, termites, wireworms, cutworms etc.
2. Application of systemic insecticides to soil or seed for protection against soil pests and sucking pests.
3. Falling of some of the pesticides on ground by run off or spraying on the crops. Moreover, it has been calculated that 50% of the sprays applied to crops miss the target and fall on the soil surface.
4. Incorporation of the remains of the treated plants with pesticides into soil
5. Pesticides may reach the untreated soil through surface run off and wind borne dust particles.

### Exposure through Water Pollution

Pesticides may get into water in several ways.

Intentional application for pest control- Pesticides may be applied over water surfaces for pest control. Some examples are as follows

1. use of copper sulfate for preventing growth of algae
2. application of weedicides in water courses and water channels to control weeds which otherwise choke these
3. use of pesticides for preventing breeding of mosquitoes
4. use of pesticides to kill and catch fishes.

Unintentional application during pest control- A lot of pesticides get into water through one of the following ways, even through we do not apply it intentionally

1. accidental application over water bodies
2. drift into ponds, lakes, streams and river water
3. discharging surplus formulation after spraying operations into rivers, ponds and lakes
4. pouring the washing waters of spraying equipments into rivers, ponds

and lakes

5. run off and erosion from treated soils.

### Exposure through Air

Air pollution with pesticides may occur by the following ways:

1. Volatilization of chemicals during spraying operations.
2. Volatilization of chemicals from treated surfaces especially by co-distillation with water vapour.
3. Small particles of pesticides remain air borne which spread and drift over large areas. Their hazard is greater when finer particle size is used as in ULV application and in case of aerial application.

### Exposure through Food

Edible portions of plants may get contaminated by the following means.

1. Foliar application of pesticides for pest control.
2. translocation from pesticides used for seed treatment.
3. drift from foliar application on adjoining crops.
4. translocation of pesticides from soil in food crops.
5. contamination of fish, grown in water contaminated with pesticides

### Occupational Exposure

During the process of manufacture of pesticides formulations packing and application of pesticides, a large number of workers get exposed to pesticides. Reports have suggested that intoxication and human infertilities from pesticides could occur most quickly in many workers when they totally neglect to follow the protective measures.

## ENVIRONMENTAL EFFECTS OF PESTICIDES

1. **Economic Aspects:** Chemical contamination of food is of undoubted economic importance for countries that depend on the export of food for a large part of their national income, as more and more countries institute control systems to prevent the import of contaminated food, it becomes increasingly important for the exporting countries to check that their products meet the requirements laid down by the prespective importers. If the food does not meet these requirements, it is usually refused entry or after appropriate treatment admitted as animal feed or for industrial purposes, thus commanding or lower price. This means that any country that does not monitor its exports for contaminants, runs a grave risk of having them rejected and of suffering heavy economic losses in consequence.
2. **Development of new serious pest problems and resistence:** Pests trend to live in balance with their natural enemies and their numbers are kept at least paritialy in check. It has been a common experience

that when pest control by chemicals is introduced to a crop and or area, that spectacular increase in yields are often obtainted for several reasons. Gradually, these increases become harder to maintain even when new pesticides are used. Usually this is because few pesticides are relatively specific to pests and not only the pest is killed but natural enemies become gradually depleted. Additionally, an insect species which may cause no damage because, it is kept completely in check by predators and parasites, can become a serious new pest. Finally, frequent use can cause development of resistance to one or more of the pesticides commonly used. Thus, a dependence on continued use of pesticides has been created.

3. **Hazards to non target species and wild life:** Pesticides are toxic by their nature. They are toxic not only to the insects, pathogen, weeds and other pests, but also to non target species like animals, fish and wild life pollinators and biocontrol agents. Soil microorganisms which cause breakdown of cellulose, nitrification, turn-over of organic matter and other biological materials may also be adversely influenced by pesticides. Birds and mammal population in agro-ecosystems and natural ecosystems can suffer from exposure to pesticides. The efects include death from exposures, pest resurgence, to use of high pesticide dosages, and reduced survival growth and reproduction from exposure to sub lethal levels.

There are reports suggesting that pesticide contamination can cause egg shell thinning, increasing the chances of egg breakage, higher mortality at hatching and congenital abnormalities of the chicks leading to extinction of species in wild life ecosystem.

## EFFECTS OF PESTICIDES ON MAN

Human pesticide poisonings are major concern in producing and using pesticides in agricultural and public health programmes. It has been estimated that annually in the world there are about 7,50,000 reported pesticide poisonings cases with about 1,38,00 deaths. Some of the most serious effects of pesticides are because of the potential of accidental exposure to lethal effects of these chemicals. It has been estimated that pesticide poisoning accounts for about 3% of all accidental poisoning by chemicals. Such accidents can be to the applicator of pesticide and can come through during spraying of pesticide or other treatment operations. Other types of accident which may involve larger numbers of people are those due to unsafe packing and leakage of pesticides in storage and transport. On a number of occasions food has been contaminated in this way.

Other accidents have occured when insecticides that have been found to be effective against one type of pest have been incorrectly used for treatment against others. In the tropics, containers that have contained pesticide concentrates are attractive for household use, if such containers are used for carrying of cooking consumables, poisoning may occour. Poisoning may occur to workers who spend a large amount of time in the field. It is an increased and convenient means of a suicide, since relatively small quantities of such chemicals are usualy sufficient to kill human beings. Pesticide residues or their metabolites in food could affect man's health. Acute over exposure to pesticide can lead to acute effects on human beings and even death. Prolonged exposure to some organochlorine insecticides can lead to concentration of pesticides in tissues. Prolonged Organophosphates exposure may lead to neurological changes. Over exposure to organomercurials may cause irreparable brain damage and hexachlorobenzene exposure may cause severe skin manifestations with photosensitivity. The nematicide dibromo chlolropropane can interfere with reproductive functions and other pesticides can at high exposure, cause tetragenic effects.

## PESTICIDE RESIDUES-PROBLEMS AND MANAGEMENT

### Good Agricultural Practice (GAP) in the Use of Pesticides

Good agricultural practice in the use of pesticides is the officially recommended or authorized use of pesticides, under practical conditions, at any stage of production, storage, transport, distribution and processing of food, agricultural commodities, and animal feed, bearing in mind the variations in requirements within and between regions. GAP also takes into account the minimum quantity required to achieve adequate control, applied in such a manner that the residue left is the smallest amount practicable. Thus, GAP ensures the attainment of the desired degree of control of pests at an economic cost and with a minimum danger to operators, agricultural workers, consumers, beneficial environmental animals and environment. As a part of GAP concerte recommendations on waiting periods of pesticide should be made regarding choice of pesticides, the formulations, dosage, application, timing of treatment, and post treatment of pesticide. Implementation of all aspects of GAP in the use of pesticides will ensure the safety of the food supply.

It has been observed that in case of several widely used pesticides in India, the waiting period of pesticides under good agricultural practices after final spraying and before harvesting have been omitted on the labels and literature. This can be a hazard to human health, as the terminal residues in the harvest produce may be significantly above the maximum residues limits when no waiting period is prescribed or followed. This is specialy true for vegetables, fruits, and other crops which are harvested continously. The present state of affairs as regard to sale of pesticidal formulations without full prescription of waiting period of pesticides in India is a situation which needs to be corrected on priority. The available data on waiting period of pesticides in this regard should be utilized to ensure safety to the consumers. In case more residue data for waiting period of pesticides are required to be generated, this should be done under good agricultural practice on time bound basis, and the pesticides without waiting period should not be sold in the interest of public protection and environmental safety.

## CONTROL OF DISCHARGE OF TOXIC CHEMICALS FROM PESTICIDE INDUSTRY IN THE ENVIRONMENT

The discharge of toxic chemicals from the pesticide industry into air, water, and soil have at times, led to gross contamination of the food. In the absence of complete data on the health and ecological effects of the vast majority of pesticides in use, we can only speculate about the long term consequences. But, isolated cases of contamination and recent epedemiological studies give clear casue for concern.

India has an estimated 4,000 chemical factories. Between 1970 and 1990, production of pesticides has increased 15 fold. The production of dyes and pigments have more than doubled. These and other pesticide industries generate large quantities of toxic waste chemicals and most of it goes directly to rivers and streams with litle or no treatment, which results in the contamination of soil and ground water. Most of the industrial waste materials are decomposed by

1. thermal decomposition
2. chemical neutralization
3. land filing of solid waste and
4. biological or natural degradation.

In the thermal decomposition process, exposure of the pesticide to a high temperature (900-1000°C) for varying period of time is resorted to. This method degrades more than 98 percent of the commercial pesticide formulations. Incineration of mercury, arsenic, lead and similar acute substances is avoided unless special residue handling and disposal facilities are available. The chemical neutralization treatment is intended for organophosphate and carbamate pesticide, but not for chlorinated hydrocarbons. Nitric acid, sulphuric acid, ammonium hydroxide, caustic soda or peroxides are used to destroy the pesticides chemically. The method of land filling of solid waste is susceptible to contamination of water and ground water streams from leaching and run off of pesticides. Shallow burial with 45 cm of earth is generally accepted for small quantities of pesticides in clay soils. However, the location should be well above ground water level and several hundred meters away from the any source of water suply. The burial of pesticides in sandy soil is not recommended as it has high possibility of leaching into surface water or groundwater. The biological method of decomposition is too slow for more persistent chemicals and hence, this method is applicable only for pesticides having short half lives.

## Botanicals

The use of botanical pesticides, especially neem can play an important tool in pest management programmes. However, there is urgent need to develop cheap and simple methods for isolation of pure compounds such as azadirachtin and for development of stable formulations. Moreover, the research work has been carried on crude extracts of neem and there is need to isolate pure chemicals responsible for insecticidal, fungicidal, nematicidal and nitrification inhibitors. In the light of the rapid progress over the past decade, the use of such methods can be expected to grow dramatically. In particular, there is a need for more sophisticated delivery systems derived through advances in biotechnology. Biotechnology will play a major role in developing new approches to pest control. The development of genetically engineered microbial pesticides and development of transgenic crop plants with insect resistant genes from *Bacillus thuringiensis* are exciting areas of research of pest control. Biotechnology has also played an important role in the development of breeding of pest resistant cultivars and special emphasis should be given to biotechnological approach to exploit some clonal variation and genetic engineering to evolve new multiple resistant varieties.

## Biorationals

The term biorationals has been carried to cover a range of alternative approaches to synthetic pesticides. They are generally of greatly increased specificity and therefore avoid non-target mortality and the associated problems. Insect pathogens are specifically playing an increasingly important role in crop protection. The bacterium *Bacillus thuringensis* (Bt) which produces insecticidal crystal proteins, has been extensively used for control of lepidoptera, diptera and coleopetera pests. So far different crystal proteins and there corresponding gene have been identified. Viruses, bacteria and nematodes, all have great potential as pest control agents, and methods for their efficient and low cost production have been developed. A number of commercial mycoherbicides are available in USA including colleg, a formulation of the plant pathogen *Colletotrichum gloesporioides* for the control of weed species *Aeschynomene virginica*. A number of natural and synthetic harmonal chemicals known as insect growth regulators interfere with harmonal regulated processes of insects. Insect pheromones and kairomones offer great promise in pest management. While pheromones cause mating disruption of insects, the kairomones can be used to increase the efficiency of parasitoids and predators. Research efforts would have to be directed to evaluation of physical, chemical and biological factors influencing pheromonal catches for development of high efficiency pheromonal trips and formulations.

## Integrated Pest Management Approach

Connected with the judicious and need based use of pesticides is the subject of Integerated Pest Management, commonly known as IPM. This is a multidimentional technology which has tremendous benefits for pest control and one of its big pest advantage is in the minimized use of chemical pesticides. Integrated Pest Management has as many definitions, as practioners, but there is consensus on its major feature, minimal use of pesticides and maximum reliance on natural regulatory mechanisms to maintain pests below the level at which they cause economic damage. In most of the crops, it is not known that what population density will cause crop loss. Serious efforts should be made to workout economic thresholds, the population levels at which economic loss is inflicted. If threshold levels can be predicted then timely action (Control action) can be taken to cut crop loses. For developing economic thresholds, data should be generated through survey and survelliance programme which in turn will require effective sampling techniques.

Under IPM more emphasis is laid on:

1. Regular pest survelliance for need based and timely application of effective pesticides instead of calender based prophylactic measures using broad spectrum pesticides.
2. Conservation and augmentation of beneficial species such as parasites, predators and other micro-organisms.
3. Promotion of improved cultural practices including use of resistant varieties.
4. Use of non chemical methods of pest control
5. Use of botanicals
6. IPM demostrations in farmer's fields.

A large number of IPM centres are being proposed throughout the country in next few years to promote this technology and to limit the use of pesticides only on need based level as part of the IPM.

## Healthy Seed and Planting Material

Number of pest problems can be minimized by using high quality pest free seeds. Sowing of healthy planting material also reduces pest incidence. Creation of severe artificial epipytotic conditions has enabled screening of parental and early generation material and in the release of a number of resistant varieties. These varieties are core in IPM activity to curtail crop loses.

## Improved Methods of Application of Pesticides

There have been spectacular advances in methods of application of pesticides in developed countries consequently, with education, application of pesticides has become progressively safer. By contrast, much of the application equipment used in developing countries is poor and maintenance and supplies of spares inadequate. Pesticides are often applied with handsprayers or drawn sprayers or by dusting equipment, which are inefficient. Consequently more hazards exists to the applicator than in corresponding application methods in developed countries, due to both lack of training and climatic conditions. Spray equipment that should be used include ultra low volume sprayers, electrodyne sprayers, granule applicators, aerosols and many other methods that minimize side effects. All these techniques serve to decrease the load of pesticides in the environment.

In India, the semi automatic knapack sprayers are cheap, easily servicable and ideal for small farms. By 2000 AD, India alone would need eight million power sprayers and dusters of two HP engines. At the moment, indigenous production hardly meets fifty percent of the demand. There is scope to go in for delivery systems that are light and possessing better nozzle regulation facilities. If the droplet size is large, it would result in poor coverage, less penetration, and more chemical usage to attain the same level of control. By reducing the droplet size, using the apropriate nozzle and power sprayer for the same amount of suspension, eight times more droplets are produced with better coverage and a greater degree of pest control. Unfortunately, a majority of Indian farmers are totally ignorant to the need to have a better spray system or pesticide delivery system. These changes were in response to the demand for greater economy, precision and convenience and the need for greater safety to agricultural workers, observers and the environment.

In India, all type of pesticide sprayers are ear marked for manufacture in small sector only. The quality of machines is often variable and consequently leading to defective spray machines which help to aggrevate the dangers of dermal and inhalation exposure. It is suggested that BIS mark certificates quality system may be made compulsory. It would ensure that only standard and safe application equipment will be available to the farmers.

## Weather Based Prediction and Agronomic Implications

The population build up of insect pests and pathogens depends on favourable weather requirements of temperature, rainfall, leaf widthness and other parameters. By monitoring weather, forecasts are made and farmers are advised to spray or not to spray their crops. Prediction based spray schedule substantially reduces the number of sprays needed and adds to greater profits. Prediction system and computer models have been developed and tested for wheat rusts, potato late blight and others. For plantation crops and commercial crop models can be of great use.

Agronomic adjustments help in avoidance of pests by modifying the microclimate. For example, flooding the banana field for quite some time prior to fresh planting, minimizes the severity caused by *Fusarium oxysporum*. Against most of the soil borne diseases, organic amendments reduces the disease severity.

## Selective Pesticides

An important characteristic of insects is the presence of chitin in the cuticle. Any interference in the synthesis of chitin would prove fatal to insects. Indeed recently, a number of analogues of benzoyl phenyl-urea, interfering in the deposition of cuticle by inhibiting the synthesis of chitin would prove fatal to insects. Investigations are also continuing to explore other areas of insect physiology and biochemistry which lend themselves to making selective chemicals for insect control. In coming years, it may be stated that insect control will reach a stage of sophistication, which would enable one to control a particular insect pest species without harming others. These chemicals if used, will be easily biodegradable and will also not have the undesirable effects on the environment. It is also certain that this aim will be achieved only by a sound knowledge of basic researches in insect biochemistry and physiology.

## Training

Training on safe and judicious use of pesticides should be imparted to the farmers by the extension functionaries of the Government of India and ICAR institutes and the Agricultural Universities. Much more training at Block level would increase awareness among the farmers. But, this has not been properly regulated or monitored. Further, monitoring of such activity will help in overcoming the deficiencies. Supply of a audio-visual aids, demonstration kits to the extension functionaries will enable them to impart training more effectively. Pesticide dealers selling pesticides play an important part in educating the farmers. There are about 92,000 dealers in India in the public, co-operative and private sectors. If they play their part properly in educating the farmers, about safe and judicious usage of pesticides and in ensuring good quality of products sold by them, then a substantial improvement can be brought about in present ground level situation.

The role of voluntary organisation like voluntary health association, Farmer's Forum and Pesticide Associations of Pesticides is also quite important in this regard. They may also consider achieving this goal by going ahead with missionary spirit. The educational syllabus of both agriculture and medicine for graduates and post graduates do not have the right emphasis on safe and judicious usage. The course should be scientifically designed and taught with adequate emphasis to become meaningful. The concept of agro-medicine can easily be included at least in our agricultural graduates educational course.

There is also a woeful lack of course on pesticides toxicology for health professionals in India. There is need to introduce adequate courses in this direction so that the medical and para-medical personnel are able to handle accidents with confidence and if referred, guide the users to the right degree of precautions needed in case of poisoining by pesticides and its prevention.

## PESTICIDE REGULATIONS AND THEIR ENFORCEMENT IN INDIA

Most of the constituents of the environment e.g., land, water, forests etc are under the purview of the states in India. For a variety of reasons, principally financial, the states are not usually in a position to take effective action against pollution or in promoting environmentally sound management strategies for conservation and development of these resources. It has thus become necessary for the union government to intervene when the said sources are threatened with serious depletion. The union government utilized their provision to pass various acts in order to protect the population, present as well as the future generations and environment from what appears to be an invorable move towards self destruction. The following acts have been passed by Government of India:

## INSECTICIDE ACT

The pesticides are basically toxic chemicals and their use is inherently accompanied by hazards and ecological consequences. The Government of India passed the Insecticide Act in 1968 and this was enforced from 1971 to regulate import, manufacture, transport, distribution and use of insecticdes with a view to preventing risk to human beings and animals.

### The Wildlife (Protection) Act

The wildlife (Protection) act governs wildlife conservation and protection of endangered species. The Act prohibits trade in more endangered species. The centre provides financial assistance to states for-

1. strengthening management and protection of infrastructure of national parks and sancturies
2. protection of wildlife and control of catching and illegal trade in wildlife products
3. captive breeding programmes for endangered species of wildlife
4. wildlife education and interpretation and development of selected zoos.

### The Water (Prevention and Control of Pollution) Act

The Water Act defines water pollution, prescribes penalties and establishes an administrative machinery called the water pollution. Boards are set up at the central and state level in order to control and prevent pollution of water. The coverage of act is quite comprehesive in that it includes streams,

water river, water courses, inland waters, subterranean waters, and tidal waters under state jurisdiction. The Water Act prohibits dumping of poisonous, noxious or polluting matter into streams and wells as well as any activity which impedes the proper flow of stream causing aggravation of pollution due to other issues. The act subjects the discharge of sewage or trade effluents into streams and wells to the prior consent of the Boards.

### The Air (Prevention and Control of Pollution) Act

The Air Act was passed in 1981 to mainly regulate and control emissions from automobiles and industrial plants. The Central Boards for Prevention and Control of water pollution is authorised to implement and enforce the act also. This body lays down standards for quality of air. After consultation with the state Boards, the state government may declare any areas within the state as air pollution control area and protects the use of any fuel other than approved fuel in the area causing air pollution. Further, no person shall, wihout the previous consent of the state board operate any industrial plant involving industries specified in the schedule in air pollution control area.

### The Environment (Protection) Act

The Environment (Protection) Act was passed by the parliament on 23rd May, 1986. By virtue of this Act, the Union Government has armed itself with considerable powers deemed necessary for the prevention, control and abatement of environmental pollution. The process include co-ordination of action by states, planning and execution of nationwide programmes, laydown of environmental quality standards, specially those governing emission of discharge of environmental pollutants, placing restrictions on the location of industries, handling of hazardous substances, prevention of enviornmental accidents, research, inspection of polluting units and establishment of laboratories and dissemination of information.

## PLANT PROTECTION RESEARCH AT ICAR

Plant protection research is an integral part of all the crop based institutes and crop improvement projects. The All India Co-ordinated Research Project on diffeerent aspects were started to fill up the lacunae in research and to strengthen the efforts at national level. The co-ordinated Research Projects viz., biological control of crop pests, rodent management, nematodes control, honey bee research, seed borne diseases, white grubs management, economic ornithology and pesticide residues. In operation during the VII plan period, apart from the existing projects, three new projects i.e. Betelvine diseases, Agricultural acarology and apple scab were put into operation. The main thrust and achievements in various All India Co-ordinated Research Project in plant protection is envisaged as follows.

## Biological Control of Insect Pests and Plant Pathogens

- **Dissemination of biological control agents:** Use of novel methods such as insect growth, regulators, pheromones and kairomones etc.
- **Seed borne diseases:** Standardisation of procedures for producing disease free seeds.
- **Betelvine:** Management of bajras and production of stocks and cuttings free of diseases and pests.
- **Apple scab:** Disease management of scab disease and development of a forewarning system.
- **Agricultural acarology:** Management of key mite pests of important crops.
- **Honey bees:** Emphasis on bee breeding and bee pathology.
- **Economic ornithology:** Management of predatory birds, herd conservation for natural control of pests.
- **Rodent control:** Integrated control approach of rodents and promotion of social engineering concept.
- **White grubs:** Management of white grubs by chemical and by non-chemical methods.
- **Nematodes:** Resistant varieties to nematodes and use of cultural and chemical control of nematodes.
- **Pesticide Residues:** Waitng period of pesticides on crops.

***National centre for integrated pest management to develop and promote IPM technologies for major crops to sustain higher crop yields with minimum ecological implications.***

# CHAPTER 17

# Toxicology and Safe Use of Pesticides

## FUNDAMENTALS OF TOXICOLOGY

Toxicology is a branch of the science studying the properties of the pesticides used in agriculture, their action on warm blooded animals, insects, bacteria, fungi, plants and ecological systems.

### Toxicity and Hazards

There is a marked distinction between toxicity and hazard. These two terms are not synonymous. The word toxicity refers to the inherent toxicity of a compound, in other how toxic it is to animals under experimental conditions. The word hazard refers to the risk or danger of poisoining when a chemical is used or applied. Hazard depends not only on toxicity but also on the chance of exposure to toxic amounts of the material.

1. **Acute Toxicity:** Acute toxicity of an organisms by a pesticide occurs when the pesticide acts at once. It manifests itself in upsetting of the vital activity of the organism with a possible lethal outcome. The acute toxicity is obviously of paramount interest to people engaged in manufacturing and preparing formulation of pesticides and to those responsible for application.
2. **Chronic Toxicity:** Chronic toxicity of an organism is the result of the repeated action of relatively small amounts of pesticide and manifests itself in slowly developing malfunction of normal, vital activity. The toxicity of pesticides varies, depends on their amount, the ways of penetration, the duration of action, the state of organism and the environment etc. In the early stages of developing a pesticide for further experiments and exploration, toxicity data are collected on the pure toxicant. These tests are conducted on test animals that are easy to work with and whose physiology, in some instances, is like that of human e.g. white mice, white rats, white rabbits, guinea pigs, and beagle dogs.
3. **Acute oral Toxicity:** Acute oral toxicity is that property of a substance which causes adverse effects in an organism by swallowing of that substance.
4. **Acute Dermal Toxicity:** Acute Dermal toxicity is that property of a substance which causes adverse efects in an organism if, the substance is being absorbed through the intact skin of the organism.

5. **Inhalation Toxicity:** Inhalation toxicity is that property of a substance which causes adverse effects in an organism by the substance being absorbed during respiration of the organism.
6. **Mutagenicity:** Mutagenicity is that property of a substance which induces changes in the genetic complement of somatic or germinal tissue in subsequent generations. Objectives for mutagenicity assessment is to detect the capacity of a chemical to alter the genetic material in the cell.
7. **Teratogenicity:** Teratogenicity is that property of a substance which produces or induces functional deviation of developmental anomalies in an animal embryo of foetus. This study is designed to examine the potential of test substance to induce permanent malformation and function disorder of the foetus as a result of exposure of the mother during pregnancy. Data of this study provide information on the safety for the applicator, particularly for woman in pregnancy.
8. **Oncogenicity:** Oncogenicity is the property of a substance which produces or induces either benign or maligant tumors in living animals. The purpose of this study is to examine whether, the test substance has the potential to cause carcinogenicity.
9. **Carcinogenicity:** Carcinogenicity is a specific form of oncogenicity characterised by malignant tumors.
10. **Delayed Nueropathy:** Delayed nueropathy is that property of a substance which causes acetyl chloinesterase depression or has the ability to cause it. Hens are used for this study.
11. **Reproduction Study:** Two generations reproduction test is designed to provide information concerning the general effects of the test substance on gonadal function, uterus cycle, mating behaviour, conception, parturition, lactation, weaning and growth and development of the offspring. The test species is rat.
12. **Primary Eye and Dermal Irritation Studies:** These studies serve to identify possible hazards due to exposure of the eyes and skin. These studies serve for the direction on the wearing of protective goggle and clothing. Rabbits are used for these studies.

## Measurement of Toxicity

Chronic toxicity studies are conducted on two species for extended period and the compound is usually incorporated in animals daily diet. Inhalation studies may involve any of the test animals, but rats, guinea pigs and rabbits are most commonly used. Before pesticide is released for public use, the manufacturer must declare toxicity to the rats under laboratory conditions. This toxicity is defined by the $LD_{50}$, expressed as milligram (mg) of toxicant per kilogram (kg) of body weight, the dose that kills 50 percent of the test animals to which it is administered under experimental conditions. The $LD_{50}$ value is measured in terms of oral (feed or placed directly in the stomach of rats), dermal (applied to skin of rats) and respiratory toxicity inhaled.

The $LC_{50}$ is somewhat similar concept to $LD_{50}$. It is the concentration usually in the air or water surrounding experimental animals which causes 50 percent mortality. It is therefore a convenient index for studying vapours inhaled by mammals or water soluble pesticides in context with fish. Related terms are $ED_{50}$ and $EC_{50}$ which are respectively, the 50 percent effective dose and 50 percent effective concentration and are applicable, when some criterion of toxicity other than death is adopted. The $I_{50}$ is similarly the concentration which inhibits by 50 percent of an in vitro enzyme systems.

To determine $LD_{50}$ of a substances, batch of animals are exposed to a range of doses and the percentage of killed organisms in each batch is recorded. The results can usually be represented by the sigmoid curve (a). The curve is steepest in the region of 50 percent response so that a small change in concentration in this region of the curve causes a large change in percentage kill. The dose which kills half the organism is thus a more sensitive index of toxicity than any other dose and that is why $LD_{50}$ is usually adopted as a standard for comparing the relative toxicity of the substances.

### Estimation of Toxicity to Humans

In order to assess the potential chronic toxic hazard of a pesticide, individuals of a suitable animal species are fed with varying amounts of the substances and the highest concentration having no effect on any organism in that batches of animals is determined. This is known as threshold daily dose. Numerous observations are made during the longer duration feeding trial by autopsy examination when animals are eventually sacrificed. The data includes change in body weight, activities of enzymes in serum, liver and behaviour changes, (after death) histological examination of liver, kidney, lung and brain. In order to arrive at a figure for acceptable daily intake (ADI) for man, threshold daily dose (mg/kg/day) is determined using an appropriate animal species and this is divided by safety factor. The later is often 100.

## CLASSIFICATION OF PESTICIDES BY HAZARDS

WHO (1975) has given the following criteria for classification of pesticides.

WHO criteria for classification of pesticides:

| LD50 for rat (mg/kg body weight) | | | | |
|---|---|---|---|---|
| | Oral | | Dermal | |
| | solid | liquid | solid | liquid |
| Extremely toxic | 5 or less | 20 or less | 40 or less | 10 or less |
| Highly hazardous | 5-50 | 30-200 | 10-100 | 40-400 |

| | LD50 for rat (mg/kg body weight) | | | |
|---|---|---|---|---|
| | Oral | | Dermal | |
| | solid | liquid | solid | liquid |
| Hazardous Moderately | 50-500 | 200-2000 | 100-1000 | 400-4000 |
| Slightly hazardous | over 500 | over 2000 | over 1000 | over 4000 |

The terms solids and liquids refer to the physical state of the product or formulation being classified.

## Pesticide Toxicity and its Label

All pesticide labels must contain 'signal words' in bold print' to attract the attention of the buyer/user; Danger-Poison-warning-caution.

- **Category I:** The word Danger-poison and the skull and crossbones symbol are required on the labels for highly toxic compounds. Pesticides having $LD_{50}$ range 0-50 mg/kg are grouped into this category.
- **Categroy II:** Warning is required on labels for all moderately toxic compounds. Pesticides having $LD_{50}$ of 50 to 500 are grouped into this categroy.
- **Category III:** The caution is required on labels for pesticides having $LD_{50}$ of 500 to 5000 mg/kg.

Besides, important information about the name of the pesticide, its concentration, type of formulation, date of manufacture, expiry date and Registeration number of the pesticide, antidotes is provided on label. The label contains a diamond shaped warning symbol. It is divided into two triangles by a horizontal line. The colour of the lower triangle gives on idea about the degree of toxicity or potential hazards of the pesticides.

| Colour of the lower triangle | Category | Single word in upper triangle |
|---|---|---|
| Red | Extremely toxic | Poison |
| Yellow | Highly toxic | Poison |
| Blue | Moderately toxic | Danger |
| Green | Slightly toxic | Caution |

**Information on the leaflet**: Due to limitation of space on the pesticide container, detailed information on the use of pesticide is given on the leaflet. The following information is normally available on the leaflet:

- Particular of application of pesticide on various pests
- directions for use
- Warning and cautionary statement
- Container disposal and
- Storage information.

## SAFETY WITH PESTICIDES

### Safe Storage

1. Store pesticides in their original containers.
2. Do not store pesticides near food, feed, or a source of portable water.
3. The storage area should be well ventilated. Ventilators should be located near floor level.
4. Floors should be of impervious material such as cement or stone.
5. Keep pesticides away from the reach of the children
6. Dampness from the soil should not rise into walls or floor.
7. Warning signs should be placed on the outside of all walls of the store.
8. The store should have an adequate size.
9. Check periodically the stored pesticides for signs of damage and leaks.
10. The storage site should be accessible preferably from all sides.

## SAFETY IN APPLICATION OF PESTICIDES

### Before Application

1. Read label and leaflet carefully and follow instructions there in.
2. Make sure that the application equipment is in working order
3. Never spray pesticide alone while handling highly toxic pesticide.
4. Ensure that water, soap, and towels are available at the site of operation.
5. Ensure that goggles, shoes and gloves are available before application of pesticides.

### During Spraying

1. Wear personal protective clothings
2. Spray crops with the wind. In other words, spray with the wind coming from back.
3. Ensure that there are no animals, people, food or animal feed down wind to the area i.e. in the direction in which the wind is blowing.
4. Check the sprayer and sprayer equipment for leaks.
5. Do not blow out blocked spray nozzles with the mouth.
6. Apply pesticide when needed.
7. Take care to prevent the applied pesticide from contaminating nearly streams, ponds, lakes or wells.
8. If during spraying the person feels ill or notices any irregular body symptoms, work should be stopped and medical care should be sought immediately.
9. Spray pesticides in the morning or evening. Don't spray pesticides during the noon.

### After Spraying

1. Empty the spray tank completely during spraying.

2. Never empty the spray tank into the irrigation canals water ways, ponds or wells.
3. Do not leave dirty spray equipment unattended. Immediately clean up and decontaminate.
4. Decontaminate or destroy all devices such as buckets, sticks, measuring cups etc. used in preparation of spray solution.
5. Take a bath with plenty of clean water and soap
6. Wash cloths separately
7. Mark the sprayed field and prevent unauthorized entry into the treated area.
8. Decontaminate all protective clothing and footwear.

## EMERGENCIES

Accidents, while using pesticides may occur inspite of taking all precautions. The symptoms of mild poisoning include headache, giddiness, weakness, nervousness, blurred vision, excess perspiration and eye irritation. If severe poisoning occur, high fever, stroke and loss of consciousness may also take place.

## FIRST AID

### Inhaled Poisons

1. If victim is in an enclosed space, do not go in after him, without an air supplied respirator
2. Carry patient (do not let him walk) to fresh air immediately.
3. Open all doors and windows
4. Loosen all tight clothing
5. Apply artificial respiration if breathing has stopped or is irregular.
6. Keep patient as quiet as possible.
7. If patient is convulsing, watch his breathing and protect him from falling and striking his head. Keep his chin up so his air pasage will remain free for breathing.
8. Prevent chilling (wrap patient in blankets, but don't overheat) Do not give alcohol in any form.

### Swallowed Poisons

1. The most important choice you must make when aiding a person who has swallowed a pesticide is whether, you should make the victim vomit. The decision must be made quickly and accurately, the victim's life may depend on it. Usually, it is best to get rid of the swallowed poison fast. But, there are exceptions.
2. Never induce vomiting in the victim if unconscious or in convulsions. The victim could choke to death.
3. Never induce vomiting if the victim has swallowed a corrosive poison. A

corrosive poison will burn the throat and mouth severely while coming up as it did going down.

4. Never induce vomiting if the person has swallowed petroleum products (Kerosene, gasoline, oil, lighter fluid)

## How to Induce Vomiting

Do not waste a lot of time inducing vomiting. Use it only as first aid until you can get the victim to a hospital. Make sure the victim is lying face down or kneeling forward while resting. Do not let him lie on his back because vomited matter could enter the lungs and do more damage. First give the patient large amounts of milk or water one to two cups for victims up to five years old; up to 0.9 litre for victims of five years and older.

Induce vomiting by using syrup of ipeac (use only on physician's orders) or by putting your finger or the blunt end of a spoon at the back of victim's throat. Do not use anything that is sharp or pointed. A glass of soapy water or salt water will also cause the victim to vomit.

## Dilute Poison Quickly

The best first aid for a person who has swallowed a poison is to dilute the poison as quickly as possible and to neutralise the acid or alkali causing the burns. Also, get the victim to a hospital without delay.

For acid or alkaline based pesticides, give the victim water or preferably milk-240 ml to 480 ml, for victims under five years; up to 0.9 litre for patients over five years. Milk is better than water because it dilutes and helps neutralizes the poison. Water only dilutes the poison.

- If you are sure the poison is an acid, give the victim milk of magnesia (15 ml to 24 ml of water), baking soda or chalk in water.
- If you are sure the poison is an alkali, give the patient lemon juice or vinegar.

1. **Universal Sponge:** Use these 'sponges' to absorb excess poisons only after first aid. Suggestions for the corrosive or noncorrosive poisons are followed.
2. **Activated charcoal:** It absorbs many poisons at a high rate. Mix it with water into a thick soup for the victim to drink. Activated charcoal is found in aquarium filters or is available from a drug store.
3. **Home made absorber:** A home-made 'universal sponge' for poison is a mixture of four tablespoons of toast (burnt black), two table-spoons of strong tea (instant ice tea mix will do), and two tablespoons of milk of magnesia. This is used to absorb and neutralize most poisons.

## Shock

Sometimes poisoned victims go into shock. If untreated or ignored, shock can kill victim even if, the poisoning injuries would not have been fatal.

**TABLE 15:** Antidotes for the treatment of pesticide poisoning

| Pesticide | Symptoms of Poisoning | Antidotes |
|---|---|---|
| Aldrin, chlordane, Heptachlor, DDT HCH, Lindane Dicofol endosulfan | Nervousness, headache, dizziness, nausea, vomiting, tremors, and diarrhoea | i) Phenobarbital 0.7 g per day or phenobarbital 0.5 g per day ii) Donot give morpholine, theophyline or aminophylline |
| Carbamates Aldicarb, carbaryl, carbofuron, | giddiness headache and nausea, diarrhoea blurred vision | i) Inject atropine sulphate 2-4 mg for an adult respectively for 24-48 hours. |
| propoxur | muscular twitching, convulsion | ii) Convulsions and anxiety can be treated by 5-10 mg of diazepam injected intramuscular |
| | | iii) While keeping the patient fully atropinised administer also an oxime if available e.g. 2-PAM 100-200 mg IM or IV for adults or 155 mg/kg after 1-2 hours. |
| Organophosphates (Malathion, Parathion, Acephate, Monocrotophos, Oxydemeton, phorate Phosalone, Quinalphos Oxydemeton, temephos, Phosphamidon, | Same as carbamates | Same as carbamates |
| Synthetic Pyrethroids (Cypermethrin, deltamethrin, Fenvalerate Fluvalinate, Fenpropathrin) | Irritation of oral/nasal mucosae salivation, convulsive seizurs, facial sensations | i) Perform gastric lavage followed by saline cathartic with sodium sulphate solution ii) Control seizurs with injectable diazepam or barbi-turates. |
| Mercury compounds | Tingling of figures,tongue, tips may occour, Headache, loss of peripheral vision, contact with skin may cause blisters or dermatitis | i) Inject freshly prepared100 to 200 ml of sodium formaldehyde sulphoxylate solution intervariously ii) Give 100 ml of 10 per cent calcium gluconate solution introvenously for muscular spasm |
| Fumigants, (Ethylene bromide, Methyl bromide Ethylene dibromide Carbon tetrachloride mixture | Cause respiratory irritation and itching, dermatitis blistering of skin and carneal, ulceration, dizziness, convulsions and respiratory failure | i) Administer artficial respiration ii) Administer amino phylline 1 mg/kg slowley I.V. as a loading dose, followed by 0.6 mg/kg per hour as a maintenance regime iii) Furosamide Adults 40 mg slowly I.V Children 0.5 to 1.5 mg/kg |

| Pesticide | Symptoms of Poisoning | Antidotes |
|---|---|---|
| | | slowly I.V. |
| Dinitrocompounds (Binapacryl, Dinitrophenol | i) Yellow staining skin (ii) Tremors increased respiratory | Adequate emptying of stomach and the use of activated charcoal |
| Dinoseb | rate, sweating, nausea restlessness, rapid heart beat | followed by saline, cathortic with sodium sulphate solution |
| Copper compounds | Gastrointestinal and renal manifestations. | BAL (Dimercaprol) 3-5 mg/kg body weight |
| Anticoagulants Rodenticides (Brodifacoum, Warfarin, Bromadiolone, Coumochlor, Chlorophocinone) | Nausea, Vomiting diarrhoea, bleeding from nose, gums | i) give vitamin K1 Phytonadione 15-25 mg to adults and 5-10 mg in mild cases. In severe cases give Agnamephyton 5-10 mg to adults, 1-5 mg to children, IM or IV |
| Bipyridybiums (Diquat,Paraquat) | Irritation of mouth and throat with nausea, diarrhoea, abdominal pain | i) Large doses of vitamin C and E as antioxidants. ii) Avoid oxygen therapy for first 48 hours iii) Give charcoal 50 g in 150 ml of water or 8 beaten eggs whites |
| Zinc and Aluminium Phosphides | Vomitting, caugh, liver and kidney failure | i) Give morphine for relief of abdominal pain ii) High doses of corticosteroids and blood transfusion for treatment of shock and haemorrage |
| Nicotine sulphate Arsenic compounds | Neurological manifestations Severe gastrointerinal | Phenobarbital 0.7g daily BAL (Dimercaprol) 3-5 mg/kg body weight |

Where no antidote is recommended or none is available, use some general measures to prevent absorption of the poison. Give acti vated charcoal as an absorbent in a dose of 50 g in 400 ml of water by mouth for pesticides remaining in intestinal tract. More can be given if needed. Repeat administration, if necessary. For elimination of toxic substances from the intestinal tract, use cathartic such as sodium sulphate (30 g of sodium sulphate dissolved in 250 ml of water). The quality of liquid given should not exceed 5 ml/10 kg body weight.

## BIBLIOGRAPHY

1. Dubois D.K.P. and Geiling EMK; Text book of Toxicology, Oxford University Press, 1959.
2. Mastumure F., Toxicology of Insecticide, New York, 1975.

# CHAPTER 18

# Insecticide Act

## WHY INSECTICIDE ACT

For sustaining productivity of agricultural, horticultural, and plantation crops, adopting plant protection measures against weeds, rodents, nematodes etc. assume greater importance. It is in this context that the use of plant protection chemicals popularly called pesticides have become an essential part of production technology. These pesticides are toxic substances and required to be used judiciously and with great care to safeguard the users, animals and wild life, birds and environment from hazards. It is in this background that Government of India passed in 1968 a comprehensive legislation called the Insecticide Act.

## INSECTICIDE ACT

The main objective of the Insecticide Act is to regulate the import, manufacture, safe transport, distribution and use of insecticides with a view to prevent risk to human beings, animals and for matters connected herewith. The enforcement of the Act was transfered to the Ministry of Agriculture in the year 1970 by Ministry of Health and Family Planning. All the provisions of the Insecticide Act were brought into force with effect from 1st August,1971.

It should be noted that due to legal reasons, the Act was called 'Insecticide Act' and not the 'Pesticide Act' and therefore, under Section 93 (c) of the Act the word insecticide means

1. Any substance specified in the schedule
2. Any other such substances including fungicides and the weedicides may be included after consultation with the board and Central Government.
3. Any preparation containing one or more of such substances.

### Salient Features of the Insecticide Act

The salient features of the Insecticide Act are:

1. Compulsory registration of the product at the central level and licences for manufacture, formulation and safe use at state level.
2. The inter departmental/ministerial/organization coordination is achieved by a high-level advisory board, the Central Insecticides Board with 24 members which has now been raised to 28 members, drawn

from various fields having expert knowledge of the subject.

3. Registration Committee to attend to the task of compulory registration of all pesticides and their formulations before the states grant manufacture/formulation and sale licences.
4. Establishment of the enforcement machinery like the Insecticides Analysts and Central or State governments.
5. Establishment of Central Insecticides Laboratory
6. Power to prohibit the import, manufacture and also to confiscate the stocks.

## Functions of Central Insecticides Board

1. to advise the Central Government on the manufacture of insecticides under the (Industries Development Regulation) Act, 1951.
2. to specify the use or classification of insecticides on the basis of their toxicity as well as their being suitable for aerial application.
3. to advice on tolerance limits of insecticides residues and establishment of minimum intervals between the application of insecticides and harvest in respect of various commodities.
4. to specify the shelf life of insecticides;
5. to carry out such other functions as a resupplemental, incidental and consequential.

## Registration Committee

The registration committee consists of a chairman and not more than 5 members including the Drug Controller of India and the Plant Protection Adviser to the Government of India. The main functions of this committee are as under:

1. to register insecticides after scrutinising their formulae and verify their claims made by the importers/manufacturers as regards their efficacy and safety to human beings and animals.
2. to specify the precautions to be taken against poisoning through the use of handling insecticides.
3. to carry out other incidental and consequential matters.

## Registration Procedure

Any person who desires to import or manufacture any insecticide has to obtain a registration certificate from the registration committee. However, the entire registration procedure may be classified into four steps as enumerated below:

***Inclusion of Pesticide in the Schedule of the Act:***

Any pesticide which is to be imported or manufactured in India for use has to be first included in the schedule of the act by the central insecticides board. Any person desiring any chemical to be included in the schedule is to furnish information to the central insecticides board on the following parameters:

1. Chemical identity and physicochemical properties of the insecticide, along with technical bulletin from the principal manufacturer;
2. Bioefficacy data on the product to prove that it is a potential insecticide;
3. The registration status of the product in other countries including the country of its origin;
4. The status of the insecticide as per Pesticide Manual, Pesticide Index or in any other standard pesticide bibliographical book;
5. Detailed information on mammalian toxicity of the insecticide;
6. Any supporting evidence to illustrate, if the need of the product has been substantiated by the Central/State Agricultural Research Organisation/Agricultural University.

After an insecticide has been included in the schedule, any body can make an application for registration of the same for import or for manufacture.

## Provisional Registration under Section 9(3B)

This provisional registration under section 9(3B) is granted for a period of two years only for data generation under Indian conditions. These are as under:

Data requirement for consideration of provisional registration under section 9 (3b)

1. Technical: (Import)
   a) Source
   b) Chemical composition with complete impurity profile
   c) Chemical identity of the technical
   d) Physicochemical properties of the technical
   e) Specifications
   f) Methods of analysis
   g) Analytical report from principal
   h) Shelf life claim
   i) Chemical composition on labels.
2. Formulation (Indigenous manufacturers):
   a) Chemical composition
   b) Chemical identity of the technical and other ingredients
   c) Physicochemical properties of the other ingredients
   d) Specification as per ISI format
   e) Methods of analysis
   f) Analytical test report from independent pesticide testing laboratory.
   g) Shelf life claim
   h) Chemical composition of labels.
3. Packaging and labelling

a) Proposed labels
b) Proposed leaflets
c) Manner of packaging:
   i) specification for primary package
   ii) specification for secondary package
   iii) specification for transportation packaging
d) Instructions for storage and use, including first aid and precautionary measures proposed for labelling
e) Information regarding disposal of used packages, surplus material and washing of insecticides.

4. Efficacy
   a) Effectiveness:
   b) International data on Bioeffectiveness
   c) Results from some trials under different agroclimatic conditions of India
   d) International data on translocation within the plant or animal.
5. esidues
   a) Metabolism of the pesticide in the plants, animals, soil and water, conducted under foreign conditions with the nature of metabolities and their toxicities.
   b) Residues data on crops under foreign conditions.
6. Toxicity; (Both for Technical & formulations)
   a) Acute toxicity studies
      i) Oral
      ii) Dermal
      iii) Inhalation
      iv) Primary skin irritation
      v) Irritation of mucous membrane
   b) Subacute toxicity studies:
      i) Oral
      ii) Dermal
      iii) Inhalation
   c) Information on the following:
      i) Toxicity to birds
      ii) Toxicity to fish
      iii) Toxicity to other beneficial insects
      iv) Toxicity to livestock
      v) Human toxicity data from foreign countries
      vi) Registration status in the country of origin and also in other countries of the world with information on the commercialisation of chemicals

## Regular Registration under Section 9(3) of the Act

During the period of the provisional registration or after the expiry of provisional registration, the application is supposed to come for regular registration under Section 9(3) of the Act.

Data requirement for regular registeration under Section 9(3)

1. Chemistry
   a) Source of supply
   b) Chemical composition
   c) Specifications
   d) Analytical test report
   e) Shelf life data
   f) Methods of analysis
2. Packaging and labelling
   a) Type of packaging
   b) Manner of packaging
   c) Specifications for primary package
   d) Specifications for secondary package
   e) Specifications for transportation package
   f) Manner of labelling
   g) Instructions for storage and use etc.
   h) Information regarding disposal of used packages.
3. Bioefficacy and residues data:
   a) Effectiveness
   b) Phytotoxicity
   c) Translocation within the plant or animal being treated
   d) Persistence in soil
   e) Persistence in water
   f) Persistence in plants
   g) Compatibility with other chemicals
   h) Direction concerning the dosage
   i) Time of application
   j) Type of application
   k) Equipment
   l) Manner in which the insecticide is to be used
   m) Methods of sampling for residue tests
   n) Residue analysis in food and feeding stuffs.
   o) Residue analysis in water
   p) Residue analysis in soil
   q) Residue analysis in wild life
   r) Expected residue level in edible crop and soil
   s) Workers hazards after treatment
4. Safety/toxicity data requirements
   a) Acute oral toxicity studies
   b) Acute dermal toxicity studies
   c) Acute inhalation toxicity studies

d) Primary skin irritation
e) Irritation of mucous membrane
f) Sub-acute oral toxicity studies
g) Sub-acute dermal toxicity studies
h) Sub-acute inhalation toxicity studies
i) Neurotoxicity
j) Effects of reproduction
k) Terratogenicity
l) Carcinogenicity
m) Synergism and potentiation
n) Metabolism
o) Mutagenicity
p) Toxicity to birds
q) Toxicity to fish
r) Toxicity to other beneficial insects
s) Toxicity to livestock
t) Signs of poisoning, diagnosis of poisoning, chemical
u) tests and records on poisoning cases
v) Treatment of poisoning
w) First aid measures
x) Medical treatment
y) Human Cattle Exposure studies

## Data Requirement for Repeat Registration under Section 9(4)

Under Section 9(4) of the Insecticides Act, where an insecticide has already been registered on the application of any person, any other person desiring to import or manufacture thereof shall make an application and on payment of a prescribed fee be allotted a registration number and grant a certificate of registration in respect thereof on the same conditions on which the insecticide was originally registered. In this case, the applicant need not submit all the data on biological parameters, if he has already established the chemical equivalence with the earlier registered product. However, the following normal information have to be supplied to the Registration Committee for its satisfaction that the chemical applied for is identical with that of the earlier registered one:

1. Import of technical grade material
   a) Minimum purity and the source of principal supplier alongwith associated manufacturing informations.
   b) Specification
   c) Analytical test report (from principal)
   d) Method of analysis
   e) Shelf life claim
   f) Packaging and labelling
      i) proposed labels

   ii) proposed leaflets and
   iii) manner of packaging
2. Import of formulators
   a) Complete chemical composition
   b) Specification
   c) Analytical test report (from principal)
   d) Shelf life claim
   e) Method of analysis
   f) Information regarding manner of packaging and labels and leaflets
3. Indigenous manufacture of technical grade material
   a) Chemical composition, both purity and impurities
   b) Specification
   c) Method of analysis
   d) Analytical test report
   e) Raw material used, their source of supply and the reaction steps involved during the process of manufacture
   f) Shelf life claim
   g) Packaging and labelling

# CHAPTER 19

# Pesticides Formulations

Pesticide formulation is the process of transforming pesticide chemical into a product which can be applied by practical methods to permit its effective, safe and economic use. Alternatively, it can be defined as a physical mixture of one or more biological active chemicals with inert ingredients which provides effective and economic control of pests.

***Reasons for Formulations***

1. Ease of application and handling
2. Greater effectiveness
3. Good and uniform distribution of pesticide
4. More surface area of the pesticide will be exposed if mixed with the diluent than if applied in lumps
5. Overdosing of material may prove to be phytotoxic
6. Hazardous to human beings if concentrated pesticide is applied
7. Safety to forms of life other than target organisms

## TYPES OF FORMULATIONS

### Dusting Powders

Dusts are very finely powdered dry formulation. Pesticides are formulated to field strength which may vary from 1 to 10 percent active ingredient depending upon the potency of the pesticide and rate of application.

These essentially contain a toxicant mixed intimately with an inert diluent or a carrier. A distinction must be made between diluent and carrier.

### Diluent

A diluent is one with low sorptivity, which is used to extend dust concentrate or a toxicant to a low field strength dust.

### Carrier

A carrier is a material of high sorptivity and low bulk density which is used to make dust concentrate of above 10%.

## METHODS OF PREPARATION OF DUST

Dusts are made by two methods:

*Straight Mix*

Pesticides and dust carriers are ground together in Raymond Kiln or micronizer mill.

*Impregnated Dust*

1. Moving column of carrier and over it insecticidal liquid is poured.
2. Insecticide is dissolved in solvent and carrier is saturated in it and solvent is then evaporated.

Impregnated dust is better than straight dust. From toxicological point of view, 3% impregnated dust is roughly equal to the 5% straight dust.

## CHARACTERISTICS FOR SUITABILITY OF CARRIERS AND DILUENTS

The most important dry carrier and diluents used in pesticide formulations are inorganic materials principally of natural origin. They are processed for use in pesticide formulations by many techniques ranging from simple drying, pulverizing to washing, and air floating and calcining. The properties of powdered carriers and diluent, are described below.

### Specific Gravity

The material used for dust formulation should have low specific gravity.

### Bulk Density

Desirable bulk density of 20 to 40 (lbs cft) should be used.

### Particle Size

Particle size has a tremendous effect upon the performance of dust formulations. Particles of 10μ exhibit poor performance and particles of 5μ exhibit very good performance. Adherence increases by decreasing particle size but, below 2μ size the particle adhere each other to form aggregates. Usually, dust formulations involve particle size from 10 to 100μ.

### Particle Shape

The shape of the dust particles may be spherical, polyhedral, tublar, irregular or otherwise and is responsible for many differences in dust performance. Highly irregular material tend to flow poor.

Materials made up of widely different shapes may not mix up evenly and may aggregate during application.

The finer dusts having less toxicant but more specific surface are more effective.

### Hardness

The hardness of solid particles is dependent upon the composition of sand particles and may cause serious abrassion of shaping and dusting equipment.

## Electrostatic Charge

Electrostatic charges are produced by friction between the particles themselves and between the particles and dusting equipment. The magnitude of charge depends upon the composition of the dust, the material of which the equipment is made of, the amount of agitation of pesticides, the temperature and humidity. Arsenicals, fluorides and sulphur compounds get negatively charged when injected from the equipment made of steel, iron or aluminium, which get positively charged. Gypsum and pyrophyllite also get positively charged while talc, silica etc get negatively charged.

Normally the plant surfaces are negatively charged and positively charged particles get quickly adheres to plant surfaces.

## pH

Insecticides like chlorinated hydrocarbons etc. are susceptible to alkaline pH and cannot be formulated with alkaline dusts.

## Surface Density and Sorption

Absorption and adsorption by dusts are dependent upon the nature of the surface, number of active centres per unit area, the size and the shape. Oil absorptivity is of practical importance in the field of diluents for the production of deterioration of oil containing dusts in storage, owing to caking. It is expressed as lbs of raw linseed oil for 100 lbs of powder required to produce a stiff like paste.

## Surface Acidity

Surface acidity may be defined as a disproportionate distribution of electric charges in the mineral to produce a surface with positively charged centres. These charged centres are referred to as acid sites or electrophillic centres. The acidity will in turn affect reaction with other chemicals. The strength of the acid centres may be measured by certain dyestuffs known as Hammett Indicators, which change colours at very specific acid strengths. These dyes form conjugated bases with the nucleophillic centres at different surface acidities which are defined as pK with a numerical system ranging from +7 to -8.

Consequently $pK_a$ is a measure of acid strength (i.e. the affinity for electrons) of the average acid site. Indicators used for determining surface acidity are given in Table

**TABLE 16:** Indicators for determining surface acidity

| Indicator | $pK_a$ of the indicator | Colour of natural base | Colour of conjugate acid | Weight % sulphuric acid |
|---|---|---|---|---|
| Natural red | + 6.8 | Yellow | Red | $8.10^{-8}$ |
| 4-Phenylazo- 1-naphtyl | + 4.5 | Yellow | Red | $5.10^{-5}$ |

| Indicator | $pK_a$ of the indicator | Colour of natural base | Colour of conjugate acid | Weight % sulphuric acid |
|---|---|---|---|---|
| amine | + 4.0 | | | |
| 4-Dimethyl- amino azobenzene | + 3.3 | Yellow | Red | $3.10^{-4}$ |
| 4-Phenylazo-diphenylamine | + 1.5 | Yellow | Purple | 0.002 |
| 4-Nitro-4'- nitrophenylazo-diphenylamine | + 0.4 | Orange | Violet | 0.3 |
| 1,8-Diphenyl-1,3,6,8-nona-tetraen 5-one | - 2.2 | Yellow | Red | 50 |
| Chalcone | - 5.6 | Colourless | Yellow | 70 |
| Anthraquinone | - 8.2 | Colourless | Yellow | 90 |

Pesticidal chemicals vary greatly in their susceptibility to decomposition as result of catalytic activity of acid sites. These acid sites can be deactivated with certain organic chemicals such as urea, hexamethylenetetramine, diethylglycol which preferentially share their electrons with the mineral to form a covalent bond, which is stronger than bonds formed between the pesticidal chemicals and acid centres.

## Flowability

The flowability of a powdered carrier is the rate at which the material can be poured, moved or displaced. Its significance in the formulation process is that as flowability increases, the power requirement for blending or working material decreases. This is a significant factor in field application because, as flowability increases, the flow through the dusting hopper improves, and the rate is more easily controlled.

## Dustability

Dustability is a characteristic of powdered diluents relating to the ability to flow through the air, to be transported by air currents within a limited area of application and to cling to the surface of the crop after application. There are no precise laboratory methods for testing the dustability other than practical application under actual or stimulated conditions of use.

## Abrasiveness

Abrasiveness is the property of the certain carriers and diluents to cause wear of processing equipment, metering devices and orifices of application equipment. This action significantly changes the flow characteristics of the nozzles thereby, increasing dusting volume rates. The dusting pattern is changed and resulting applications are inacurate.

Good criteria for the selection of carrier and diluent is sorptivity and surface activity followed by bulk density flowability, and capacity to absorb less moisture. The following are the main diluents used for the manufacture of insecticidal dusts in India.

- **Talc:** Talc also known as soap stone or steatite in its solid form is essentially metasilicate of magnesium ($MgSO_4.SiO_2$). Mostly talcs are of alkaline pH 8-10 and bulk density 20-40 lbs cft.
- **Prophyllite**: This is chemically a hydrous aluminium silicate ($Al_2O_3.4SiO_2H_2O$).
- **White shale**: The mineral of unknown geological class and is also used for formulation in India.

## Driftless Dust

Driftless dust was developed to resolve the drift problem of conventional dusts. It differs from the conventional dusts the mean diameter in driftless dust is 20-30 μm as compared to 10 μm of regular dust other properties are compared as follows:

| Property | DL dust | Regular dust |
|---|---|---|
| Fineness | Above 300 mesh | Above 300 mesh |
| mean diameter | 20-30 μm | 10-12 μm |
| % of particle, under 10μm | Blow 30% | About 50% |
| Bulk density | 0.7-1.1 | 0.45-0.65 |
| Floatability | Below 20 | 40-60 |

### *Advantages*

Dusts are easy to apply, less phytotoxic, require no sophisticated equipment and are useful where there is scarcity of water.

### *Disadvantages*

Transport or handling is costly, highly susceptibility to drift, less deposits due to wash of by rain, wind or gravity, Dermal toxicity is less but inhalation hazards are more, residual life is less and to control pests repeated application is necessary.

## WATER DISPERSIBLE POWDER AND WETTABLE POWDERS

Water dispersible powders are finely divided pesticidal dust concentrates blended with a wetting agent plus a dispersible agent, that will yield stable, sprayable, homogenous suspensions when diluted with water to field strength. Where as a wettable powder consists of finely ground powder of pesticide a wetting agent. The wettable powders are generally used for making paste. Normally 50% of the active ingredient present in WDP should remain suspended in water for half an hour after mixing in water. The inert material used in the manufacture of WDP is generally china clay which is hydrous aluminum silicate. A good clay has pH from 4.5 to 6.0

and iron should not more than 1% by weight.

## Criteria for WDP

1. Particle size should be less than 5 μm and all the particle should pass through 44 μm sieve.
2. It should be free flowing.
3. It should wet and disperse readily without forming lumps when diluted with water or during storage.

## Wetting Agents

The wettability and spreadiability of water dispersible powders is achieved by adding wetting and or spreading agent, which help to reduce the surface tension in water and thus aid the spray droplets to wet and then spread over the leaf surface. It is flattened on a leaf surface. The attractive force (surface tension) of the leaf tends to resist the tendency of the droplets to return to a spherical shape. If it is sufficiently large, it will overcome the cohension of the molecules of the liquid so that the later will spread. The addition of wetting agent reduces the surface tension of the spray droplet and thus aids in wetting and the spreading of the application in the pesticide formulations because of high wetting property. The wetting agents indigenously manufactured are available under various trade names such as Idet 20, Idet 10, surgaol- 0.5, Dispersol, Sarcol DNL & DNC. Wetting agents upto 2 to 3% are commonly used for WDP formulations.

## Spreadability

The angle of contact (Cos θ) of the droplet with the plant surface gives the picture of the spreading concept. The three forces namely the surface tensions of the solid (Plant surface- TVS) and liquid (drop-TLV) and interfacial tension between solid and liquid (TSI) exert force in three directions. The relationship between these three forces can be put in following equation.

$$TSV = TLV \cos(\theta) + TSI$$

$$TLV \cos(\theta) + TSI = TSV$$

$$TLV \cos(\theta) = TSV - TSI$$

$$\cos(\theta) = \frac{TSV - TSI}{TLV}$$

The angle of contact is the measure of all the three forces and lesser the angle, greater the area covered by liquid. The angle of contact is never aimed at zero nor at 180°, because of fear of dripping due to infinite spread where the coverage will be nil. An ideal angle of contact is between 20-30°.

## Dispersibility

The ability of the powdered concentrate to be suspended in water as a fine solid particle is termed as dipersibility. A fine particle size and wettability are not usually sufficient to keep the particles in suspension. The cohesive forces between the particles cause their aggregation flococultion after dispersion. To overcome this tendency to floculate, water dispersible powders contain dispersing agents which ensure that the dispersed particles will remain in suspension for a reasonable period of time. Some of materials used for dispersing agent are sodium carboxymethyl cellulose (CMC), polysaccharide gums. (eg guor gum), proteins, like animal glue, and also sodium lignin sulphonates.

## Chemical Stability

Addition of decomposition inhibitors or 'deactivators' as they are called, play a substantial role in enhancing stability of WDP. Glycols are reported to be effective stabilizers for chlorinated hydrocarbons and organophosphates. Metallic impurities like iron, catalyse decomposition of organophosphates.

## Emulsifiable Concentrates

Pesticide emulsifiable concentrates (ECs) are blends of the pesticides, solvent and emulsifier. The function of the solvent is to help in dissolving the pesticide in desired quantities and emulsifier disperses the toxicant and carrier solvent in water prior to application.

Charateristics of good solvent

1. The pesticide should be highly soluble in it.
2. It should have no phytotoxicity
3. It should have no toxicity
4. It should be easily available
5. It should have high flash point

Characteristics of good emulsifier

The emulsifying agent should

1. dissolve completely in the solution of pesticide and should not adversly affect its solubility in the solvent.
2. should confer the desired emulsification characteristics to the EC.
3. should not cause chemical decomposition of the toxicant
4. should be neutral or mildly acidic
5. be free from impurities likely to have a deleterious effect on the toxicant
6. should not contain carrier solvent which will interfere with the stability of EC.

## Types of Emulsions

Two types of emulsions are possible. Either the i) Oil in water or OW type, water is external or continuous phase and oil is the internal or disperse phase ii) water in oil or WO type, oil is the external and water is internal.

The type of emulsion generally required in spraying practice in agriculture are the OW emulsion where water in continuous phase is readily diluted with water. Inverted emulsion are not yet made in India, but their use and needs are yet to be explored.

## Function of the Emulsifier

The function of the emulsifier is to prevent separation of the pesticide dissolved in solvent from the diluent water so as to give stable emulsion. Normally when two liquids of different specific gravity such as oil and water are mixed together, they tend to form distinct layers but after agitating with emulsifier, oil breaks into small droplets that floats in water phase to give a homogenous emulsion. Each oil breaks into small droplets that floats in water phase to give a homogenous emulsion. Each droplet is surrounded by water and the emulsifying agents reduce interfacial surface tension between oil and water and thus oil droplets are prevented from coalescing. It is very necessary not to use hard water for diluting emulsifiable concentrates, since the electrolites present in water disturb the balance of forces maintaining the equilibrium between oil and water and break the emulsion into seperate components from which they are prepared.

## Emulsions

An emulsion is a heterogenous systems consisting of at least one immiscible liquid intimately dispersed in another in the form of droplets, whose diameters generally exceed 0.1μ. Emulsions essentially are unstable heterogenous systems, their stability is accentuated in suitable additives which are surface active agents.

Instability of an emulsion is first indicated by creaming, this manifests in the form of separation of the dispersed phase as a top layer containing a higher proportion of the inner phase (and consequently toxicant) then the rest of the system. As the dispersed phase progressively coaleseas, the oil droplets increase in size leading to reaggregation of an oily layer and the breakdown of emulsion in complete.

**Emulsion concentrate** (E.W. also called emulsion oil in water)

In emulsion concentrate the use of solvents has been dispersed with and it only contains an active ingredient blend of surfactant to regulate hydrophilic/lipophilic characteristics. The EW is utilized as new formulation substitutes for the emulsifiable concentrate and has the following advantages.

1. **Decreased acute toxicity:** As EW formulation does not contain organic solvents, the skin permeability, inhalation, skin and occular irritation are reduced and are confined to the characteristics of the active ingredient.
2. **Decreased vapour pressure:** In the absence of petroleum solvents, the other ingredient being less volatile vapour pressure is decreased.

3. **Higher Flash Point:** There is no ingredient in the formulation which gives off inflammable vapours. The water diluents viz water and emulsifier does not fall in this hazardous category.
4. **Less manufacturing hazard:** The inflammable solvents are not involved in formulation and manufacture of the product becomes much safer and the contamination of the skin, if any, will be of less concern.
5. **Cleaner working environment:** Less vapour pressure, decreased odour, decreased spillage of solvent system involved in the manufacture of the product, all of which contribute to maintain a clean ambient working environment in the formulation premises.
6. **Nonphytotoxic:** The EW products are not phytotoxic in nature because they do not contain the solvents.

## GRANULAR FORMULATIONS

Granular formulation is one in which the pesticide is impregnated in suitable carrier and then converted into granules.

**Types of granules:** There are two types of granules.

Extruded granules and Non Extruded granules. The extruded granules disintegrate when comes in contact with water. Non extruded granules will disintegrate after a certain lapse of time. These two types are further subdivided according to volatile matter content. LVM indicates low volatile material content and RVM indicates regular volatile material content.

Typical physical properties of these two subgranules are be as follows:

| | LVM | RVM |
|---|---|---|
| Moisture | 1% | 6% |
| Volatile matter | 5% | 15% |
| Oil absorption | 120% | 95% |
| Water retention | 180% | 135% |
| Bulk density | 30-35 lbs | 40-45 lbs |
| Specific heat | 0.25 | 0.25 |
| pH | 7-9 | 5-6.5 |

**Mean size:** Granules are available in several sizes as per requirements 10/20, 20/40, 30/60. Size 30/60 means that most of the granules will pass through a 30 mesh sieve but will be retained on 60 mesh sieve. The selection of meshing for a given application is only influenced by area requirements, application method, drift problems. Absorption increases with decreasing particle mesh size. By and large, the dominant criterion is the toxicant doses to be applied per acre and number of granules required per square feet. The approximate number of granules required per square feet resulting from application of 100 lb per acre of pesticidal granules is as follows:

| Mesh size | Particle size |
|---|---|
| 10/20 | 750 |
| 20/40 | 15,200 |
| 10/60 | 20,800 |

## SELECTION OF INERT CARRIERS

The carriers for granules are minerals like attapulgite, kaolin, alumina, bentonite etc. It may, however, be noted that efficacy of granules of some pesticide concentration on different basis differs greatly in the control of pests. For each application, choice of the carrier has to be first tested for its efficacy by biological assay. Mineral clays from which either the granules are made have acid sites in their molecules. These acid sites act as catalyst for the rapid decomposition of the pesticidal ingredients.

## TECHNIQUES OF FORMULATION

### Impregation of Unextruded Granules

A known quantity of unextruded granules is placed in a Ribbon blender so that loses by attrition are reduced to the minimum. First these granules are deactivated by a suitable deactivator. Thereafter, a solution of the active ingredient is sprayed by suitable nozzles on the granules and tumbling continued for, at least for half an hour. The granules so formed should be left for aging to enable the solvent to evaporate, otherwise it may cause caking or prove phytotoxic to the crop because of the excessive solvent adhering to it.

### Extruded Pesticidal Granules

For extruded granules, the clays are first ground with a suitable deactivator and after it is determined that the deactivation is complete, the active ingredient is added and the result of the required strength is made. The dust is then mixed with water into a kneading machine from where it is passed into a granulator, which turns the material into granules. The granules are then passed over a conveyor where hot filled gases are being let in, to enable the material to contain the minimum moisture before it is packed into drums.

### The Quality Control

The following tests must be carried out for each batch by following standard sampling methods.

- Active ingredient content
- ***Bulk density:*** The weight per cft determined for any mesh size, must be constant.
- ***Deactivation:*** Neutralization of catalytic activity is tested so that $pK_a$ value is higher than 1.5.

- *Moisture content:* Same is tested for deterring the loss in weight after 24 hours placement in an oven at 100 °C. The percentage should not exceed 6%.
- *Attrition Quality:* The attrition should not be more than 2%.
- *Grain size:* The sieve shaker fitted with the size of the specified sieves is used to test for the required grain sizes.

*Advantages*

1. These can be directly applied in soil
2. No costly equipment is needed
3. No drift problem.

## Encapsulated Granules

The principal objective of this new approach is to provide a controlled released rate of toxicant for specific pesticidal action. Microencapsulation is a process designed to apply thin, continuous, polymeric coatings on to small particulate solids, droplets of pure liquids. The factors to be considered in selecting coating material are

The chemical inertness of the material towards active ingradient.

The ability of the material to dissolve or disintegrate at a controlled rate by the action of certain environmental factors such as soil and microorganisms.

The advantages claimed to be derived from controlled release of pesiticides are:

1. Lower mammalian toxicity of the formulation.
2. Substantially extended duration of activity at equal level of active ingredient, hence better cost benefit ratio.
3. Reduction of evaporative losses.
4. Decreased phytotoxicity.
5. Reduced leaching into soil and water.
6. Regulates release of active ingredient.

## Floating Granules

Floating granules are formed from divided solid materials that are compressed into granules during compounding and manufacturing process. When floating granules are applied to field, they sink once to the bottom of the water and soon come up to the water surface and the carriers gradually dissolve into the water releasing effective ingredients from the granules. In 24 hours after application, 80% of the applied granules will be dispersed in water. Active ingredient partially comes up to the water surface in the form of oil film and when these granules are applied in paddy field the pesticide partially adheres to columns of paddy rice and partially comes in contact with rice water weevil. The ingredients spreading over the water surface, gradually sink to the bottom, forming an active layer on the surface of the soil of paddy field and passing through the soil to reach the larvae of the weevil. Such floating granule is suitable for control of rice water weevil and is now attracting public attention as a new granular formulation. The particular advantages of floating of granules include a high pesticide concentration per unit weight of the product, relatively dust free formulation and a comparatively safe package.

## Water Dispersible Granules

Water dispersible granules are granules which disperse or dissolve readily and completely in water. They provide spray slurries atleast equivalent to slurries from WP. A WDG formulation consists of hard, uniform in size, free flowing, low dust granules which readily disintegrate in water and under minimal agitation, forms a homogenous sprayable suspension. To the user they provide advantages of safety and convenience.

1. Free flowing and packages are easily, quickly and completely emptied into the spray tank.
2. There is insignificant dusting during handling and preparation of spray slurries, so that sprays can be prepared without hazard to workers or nearby areas, even when there is appreciable wind.
3. The granules are generally spheroidal, so they do not cling to dry skin or clothing in case of spill.
4. Spilled material can be readily swept up.
5. Emptied containers contain insignificant residues, thus minimising waste or hazards during container disposal.
6. Package contents can usually be measured volumetrically. The bulk density of dry WDG is not much affected by handling i.e. the loose and bulk density are very closely the same.
7. They provide rapid wetting and dispersion in the spray tank which is often better, faster and more complete than can be achieved by WP.
8. Reduced transport and warehousing cost
9. Eliminates physical problems of flowables i.e. setting, thickening etc.
10. Novel, generates consumer interest.

## Water Miscible Liquids

The technical grade material may be initially water miscible or it may be alcohol miscible which is formulated with an alcohol to become water miscible. These formulation resemble the emulsifiable concentrates in viscosity and colour, but do not become milky when diluted with water.

## Fogging Concentrates

Fogging concentrates are the formulations sold for public health purposes. Fogging machines generate droplets, whose diameters are usually less than 10 microns, but usually greater than 1 micron. The thermal fogging device utilizes a flash heating of the oil solvent to produce a visible vapour or smoke. The eluent fogger atomizes a tiny jet of liquid in a venturi tube, through which passes an ultra high velocity air stream. The materials used in fogging machines depend on the type of fogger. Thermal devices use oil only, whereas ambient generators use water, emulsions or oils.

### Baits

Baits consist of the technical pesticide compound with suitable diluent such as corn, starch, or sugar. Local baits may incorporate attractant substance or colour in a manner found to be attractive to the pest. The disadvantages of baits are that

1. It can be eaten by domestic animals.
2. It leads to the development of bait shyness, especially when dead animals are left near a bait.

### Aerosols

It is possible by suitable means, to produce suspensions of solids and liquid in air, the individual particles of which are of colloidals dimensions. Such colloidal suspensions are called aerosols. The most common method of producing aerosols is by means of release of compressed gas, with which the toxicant is mixed. The drops of aerosols are in the range 10-50μ diameter. Fluorocarbons such as diflurodichloromethane (Feron 12) and 1,1 and 1,2-dichloroetetrafluroethanes are most common propellents.

### Smokes

Pesticide is mixed with an oxidant e.g. (sodium chlorate) and combustible material (solid carbohydrate) with a retarding agent ($NH_4Cl$) in the generator and ignited, when a large amount of hot gases with high velocity are emitted from the generator causing the pesticide to vapourise and mix in air to produce a fine smoke. The period of high temperature is so short that breakdown of the active ingredient is minimised. These are used in isolated places. A special form of smoke generator is mosquito coil made from wood dust, starch, various additives and green colour and pesticide (natural pyrethrin/allethrin etc.). Effects of smokes from a mosquito coil in sequence are deterency, repellency, intereference with host location, knockdown and eventually death.

## ULTRA LOW VOLUME FORMULATION (LOW VOLUME SPRAYS)

This type of formulation is concentrated solution of the toxicant in a suitable solvent. An ideal ULV solvent is that which has good dissolving power, low volatility low viscosity, no phytotoxicity and compatible with a range of chemicals. In addition, the solvent should not cause any deterimental effect. The solvents used are Riselle oil, a mixture of heavy aromation naptha, dutrex, vegetable oils, aromatic hydrocarbonds, glycols and mineral oils. This technique has proved extremely useful where insect control is observed over vast areas.

## CONTROLLED RELEASE FORMULATIONS

Controlled release technology can be defined as a technique or method in which the active ingredient is made available to a specified target, at concentration and duration designed to accomplish an intended effect. Controlled release formulations have several advantages over conventional formulations.

1. Satisfactory effect for longer period of time
2. Small dosage
3. Longer application interval
4. Decrease in mammalian toxicity
5. Decrease in phytotoxicity
6. Decrease in environmental degradation
7. Decrease in environmental pollution
8. Decrease in leaching and evaporation
9. Decrease in reactivity
10. Changing liquid into solid
11. Masking of odour
12. Easier handling

### Kind of Controlled Release Technology

Controlled release technology is mainly attained by interacting pesticides with polymers. Several methods have been developed and are classified according to the mechanism or material used as follows:

Diffusion of pesticides in polymeric medium

### Monolithic System

Pesticides are dissolved or dispersed uniformly in polymeric or elastomeric materials

1. **Reservoir system:** pesticides are covered by membranes. Reservoir layer is mainly made of flexible PVC
2. **Polymeric pesticide:** Pesticide is bound via covalent bond to polymers. Pesticides, that contain very reactive functional groups such as carboxyl, hydroxyl, sulphydryl and amino groups can react with monomers or polymers directly. If pesticides do not have any functional groups, some bridging agents are needed in order to bind pesticides to monomers or polymers. Pesticide is released from the polymeric pesticide by hydrolysis of the polymer pesticide bond in the environment.

## MATERIALS USED FOR CONTROLLED RELEASE FORMULATION

Various kinds of polymers are used for controlled release formulations.

They should satisfy the following conditions in general.

1. The polymer should have proper molecular weight and molecular structure in order to get proper diffusion and release rates of the pesticides.
2. The polymer should not be reactive with pesticides.
3. The polymer and its degradation products must not cause any environmental impact.
4. The polymer should be stable during storage and usages.
5. The polymer should not be expensive
6. For agricultural application, it should be biodegradable to avoid environmental pollution.

Examples of polymers used for controlled release formulations are:

1. Cellulose
2. starch,
3. sawdust
4. bark
5. starch xanthate
6. starch borate
7. polyacetic acid
8. polyglycolic acid.

## ADJUVANTS

Adjuvants are materials that are added to a pesticide mixture to improve mixing and application or enhance pesticide performance. An adjuvant customizes the formulation to meet specific needs or compensate for local conditions.

Adjuvants can be used to improve the wetting ability of a spray solution, control evaporation of spray droplets, improve whether inability of pesticides, increase the penetration of pesticide through the plant adjust the pH of the spray solution, improve spray droplet deposition, increase safety of the spray to target plants, correct in compatibility problems and reduce spray drift. A few of the more important types, their usage are described below.

### Penetrant Aids

There are two significant factors that affect the biological activity of a pesticide formulation, namely the penetration of the formulation through an outer protecting lipid memberance of the pest and rate of reaction of the pesticide with a vital reactive site. Penetration may be enhanced or reduced by the addition of nonionic surface active agents to the formulations. Penetrant additives should be present at a sufficient concentration that the final spray solution will contain greater than 0.5 percent by weight of the penetrant aid.

## Stabilizers and Deactivators

Many of the clays, minerals and inorganic substances used as carriers and diluents for dry formulation have the property of surface acidity which catalyses the decomposition of many pesticides. The surface acidity must be neutralized to overcome catalytic activity. Alcohols, glycols, hexamethyleneteytramine (HMT) can be used deactivators.

PH-Regulators such as organic acids and bases prevent decomposition reactions that are favoured in certain pH ranges. Certain exponidised compounds are suitable as proton traps.

## Anticaping Agents

When dust concentrates, water dispersible powders and granular formulations are prepared at the saturation point of carriers, caping occurs in storage Diatomaccous earth, and microfine silica, and attapulgite are often used as anticaping agents.

## Dry Lubricants

Dry lubricants improve the flow or the slip characteristics of the formulations. Materials useful as drylubricants in pesticide formulation are powdered graphite, soapstone talcs and certain metal stearates.

## Stickers

Stickers prevent the run off spray solutions when applied to crops. Many protective colloids, and gelling agents such as blood albumin and polyethylene polysulphide are used as stickers.

## Antidusting Agents

Antidusting agents are generally liquid substances, which cause extremely fine particles of dry materials to adhere to each other, thus making them less susceptible to drift water soluble materials such as glycerine, peteroleum oils (desel oil, motor oil viscosity SAE 30) may be used as antidusting agents.

## Antifoaming Agents

Antifoaming agents are surfactants which suppress the tendency to foam formulation by other surfactants used as emulsifiers or wetting agents when the formulation is diluted with water. Silicones are the first choice because of their chemical inertness and alliphatic alcohols of eight to ten carbon atoms in length are also used.

## Viscosity Regulators

Viscosicity plays a leading role ULV formulations, viscosity may be raised by addition of clays such as bentonite or of water soluble polymers, or it may be lowered by certain dispersing agents. Synthetic silica can cause a drastic increase in viscosity.

## Surfactants

Surfactants are chemicals which reduce the interfacial tension between immiscible liquids or between liquid and solid surfaces. A surfactant is used for management of systems containing two immisible solvents. In pesticide formulations, the surfactant characteristics of greatest interest are wetting and dispensing for water dispersible powders and emulsifying for emulsifiable concentrates. Surfactants are classified depending on the charge of the surface active moiety, usually the lengthger part of the molecule. In anionic surfactants this moiety carries a negative charge as in soap

$$C_{17}H_3CO^-_2Na^+$$

In cationic surfactants the charge is positive

$$(C_{18}H_{37})_2N^+(CH_3)_2Cl^-$$

In nonionic surfactants there is no charge on the molecule

$$C_{15}H_{31}(-OC_2H_4-)_7OH$$

In amphoteric surfactants, solubilization is provided by the presence of positive and negative charges in the molecule

$$C_{12}H_{25}N^+(CH_3)_2CH_2CO_2^-$$

In general, the hydrophobic group consists of hydrocarbon chain containing approximately 10-20 carbon atoms. This chain may be interrupted by oxygen atoms, a benzene ring, amide ester and other functional groups. Hydrophiic, solubizing groups, for anionic surfactants include carhoxylates, sulfonates, sulfates and phosphates. Cationics are solubilized by amine and ammonium groups. Ethylene oxide chains and hydroxyl groups are the solubilizing groups in nonionic surfactants. Amphoteric surfactants are solubilized by combinations of anionic and cathonic solubilizing groups.

## Anionic Surfactants

**Carboxylates:** Soaps represent the most of the commercial carboxylates. The general structure of soaps is $RCOO^-M^+$ where R is a straight hydrocarbon chain in the $C_9$-$C_{21}$ range and $M^+$ is a metal or ammonium ion.

$$C_{12}H_{25}\text{-}O^-Na^+$$

Sulphated alcohol (Dreft)

$$R_2\,CH - O - SO_2 - O^-\,Na^+$$

Sulphated alcohol (Teepol)

$$C_2H_{25} - C_6H_4 - SO_2O^-\,Na^+$$

Sulphonated hydrocarbon (Dispersols)

$$RR'\,CH\,C_6\,H_4 - SO^-_3Na^+$$

Alkyl benzenesulfonate

$$C_{13}H_{33}\ CON\ (CH_3)\ CH_3CH_2SO_3^-\ Na^+$$

Amidosulfonate (Sodium N-olcoyl - N - methyl phosphate ester aurate)

$$C_{12}\ H_{25}\ [\ -\ OCH_2\ CH_2^-)_4\ OPO(OH)\ ONa^+$$

## Cationic Surfactants

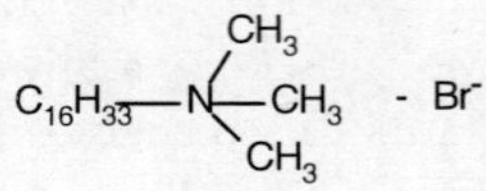

Quaternary ammonium salts (cetyl trimethyl ammonium bromide and pyridinium salts (cetyl pyridinium bromide)

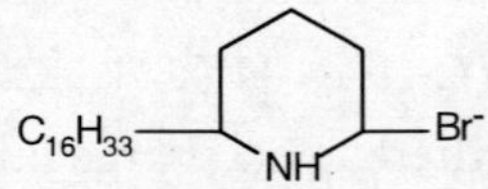

**Nonionic Surfactants:** Many non-ionics surfactants are polyethylene oxide derivatives formally derived as follows:

For example Tween 40 contains isomers of polyoxyethyl sorbition monopalmiate

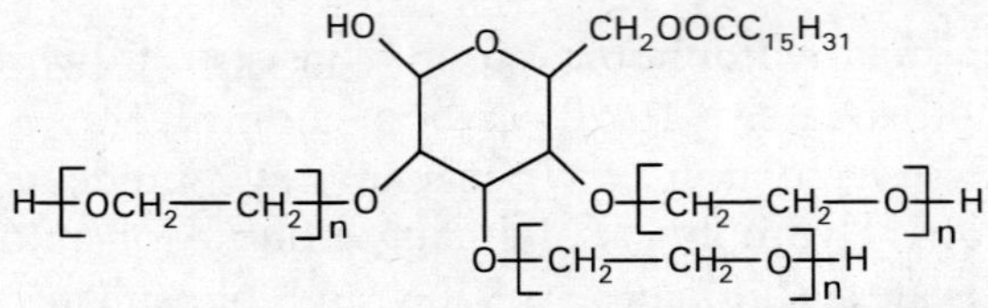

The mean value of n veries in different marketed products

## BIBLIOGRAPHY


1. Van. Valhenburg W. ed Pesticide Formulations Marcel Dehher, 1973.
2. Van. Valhenburg W., Surgeranam, B. Kheten, S.K. Pesticide Formulation United Nations Industrial Development Organisation New Age International (P) Limited Publishers 1998.